MODELING DYNAMIC PHENOMENA
IN MOLECULAR AND CELLULAR BIOLOGY

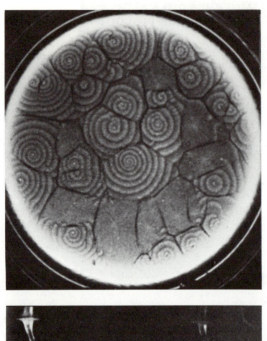

(Top) A petri dish containing a layer of slime mold amoebae. In most of the "territories" into which the dish is divided, waves of cells move inward toward a central aggregation point. See Chapter 6 for more details. [From Figure 2 of Newell and Ross (1982).] (Bottom) A slime mold aggregate in the process of constructing a fruiting body. (The time interval between photographs is approximately 1.5 hours.) Dead cellulose-filled stalk cells are ultimately topped by a lemon-shaped ball of spores. [From Plate 6 of Bonner (1967).]

Modeling dynamic phenomena in molecular and cellular biology

LEE A. SEGEL

Henry and Bertha Benson Professor of Mathematics
Weizmann Institute of Science

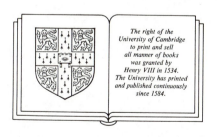

The right of the
University of Cambridge
to print and sell
all manner of books
was granted by
Henry VIII in 1534.
The University has printed
and published continuously
since 1584.

Cambridge University Press

Cambridge

London   New York   New Rochelle

Melbourne   Sydney

Published by the Press Syndicate of the University of Cambridge
The Pitt Building, Trumpington Street, Cambridge CB2 1RP
32 East 57th Street, New York, NY 10022, USA
296 Beaconsfield Parade, Middle Park, Melbourne 3206, Australia

First published 1984

Printed in the United States of America

Library of Congress Cataloging in Publication Data
Segel, Lee A.
Modeling dynamic phenomena in molecular and
cellular biology.
Includes bibliographical references and index.
1. Molecular biology – Mathematical models.
2. Cytology – Mathematical models. 3. Molecular
biology – Data processing. 4. Cytology – Data
processing. I. Title. [DNLM: 1. Models, Biological.
2. Molecular biology. 3. Models, Structural.
4. Mathematics.   QH 506 S454m]
QH506.S44   1984   574.87′0724   83–15172
ISBN 0 521 25465 5 hard covers
ISBN 0 521 27477 X paperback

Contents

Contents

In loving memory of my father,
Louis Harry Segel (1900–1983).
His confidence that we would do our best
remains an encouragement and a challenge.

Preface

This book is intended primarily for students of biology who have studied calculus for a year. Its contents have been mastered by many who have forgotten much of the mathematics they once knew; so readers should not be concerned if their calculus is rusty from disuse – the required skills will quickly come back with a little practice.

Certain mathematical techniques are imparted, but the main purpose of the book is to demonstrate that analysis of mathematical models can sometimes reveal important aspects of biological problems. This is done by means of several case studies, primarily from molecular and cellular biology. Almost all the examples concern dynamic behavior, the development with time of various phenomena.

Although it may appear simple, construction of a mathematical model is not a skill that can be learned in a semester. But in this time one *can* learn to "read" the equations of a model, and thus judge whether a theoretical paper seems promising or can be rejected at the outset as grossly deficient. Mathematical analysis of a model is another matter that can require highly sophisticated techniques, but if the reader can recognize that a theoretical attack might be of value, then he or she can doubtless obtain assistance from a friendly mathematician or physicist.

It is the recognition that is decisive. Study of the material in this book will enhance the chances that a possible application of theory will be noticed, and will help ensure that the biologist has enough in common with more theoretically minded colleagues to be able to engage in cooperative work with them.

"Art is the lie that helps us see the truth," said Picasso, and the same can be said of modeling. On seeing a Picasso sculpture of a goat, we are amazed that his caricature seems more goatlike than the real animal, and we gain a much stronger feeling for "goatness." Similarly, a good mathematical model – though distorted and hence "wrong," like any simplified representation of reality – will reveal some essential components of a complex phenomenon. The process of modeling makes one concentrate on separating the essential from the inessential. Stripped of secondary detail, a good model can permit deep analysis of interactions that are otherwise obscured. Moreover, the biological modeler may enhance understanding with the aid of various powerful concepts that have been honed in other fields. All this may suggest new and fruitful lines of experimentation.

Various aspects of modeling are illustrated in Chapter 1, a study of optimal storage policy in algae. At first, it is recommended that the student merely try to be convinced that the theoretician's line of attack is not implausible (although a more critical attitude is appropriate later). Thus, in Chapter 1, which is based on an article by Cohen and Parnas (1976), the reader should notice that the division of all cell materials into two categories, synthetic and storage, is a major simplification, but not one that is outlandish. Similarly, the assumption that both types of materials are synthesized at the same rate is quantitatively inaccurate, but this may well not mar the major conclusions. (Note that the preceding two sentences exemplify an attempt to separate the inessential from the essential.)

Once a mathematical formulation is attained as in equation (1.13), one obtains reinforcement for the view that our earlier simplifications were wise, for the simplified problem is hardly trivial. As always, if a first skeletal analysis is encouraging, one can always add more detail. (An illustration of this is the abandonment of the assumption that S_{min} is constant. Also see Exercises 1.4 and 1.7.)

The major theoretical concept introduced in Chapter 1 is that of optimization. Readers will have been exposed to this notion in studies of "maximum–minimum problems," and our application is of this type. Nonetheless, there are subtleties that usually are not exemplified in elementary mathematics courses, with their often rather sterile "applications" of the theoretical material. These subtleties hint at the fact that the role of optimization theory in biology is important but not at all straightforward. Some directions for further reading are mentioned at the end of the chapter.

The Cohen–Parnas theory suggested experiments which revealed that certain photosynthetic algae divert fewer resources into storage at higher light intensities. This contradicts the prediction of the earlier "excess theory," and it is in accord with the development of Cohen and Parnas.

It is quite typical that the experiment is *semiquantitative*. This is in marked contrast with the practice in many areas of physics, where, for example, a measurement of $198,322 \text{ cm}^{-1}$ was made for the ionization potential of helium. Divergence from a prediction of $198,298 \pm 6 \text{ cm}^{-1}$

showed that older quantized-orbit theories "led definitely to the wrong value," in contrast to the then-new quantum mechanics (Condon and Shortley, 1957, p. 347). Decisively accurate quantitative experiments usually are impossible in the biological sciences, but they are equally impossible at present in many areas of the physical sciences, such as meteorology and oceanography. Nevertheless, as the simple experiment of Cohen and Parnas exemplifies, theories need not be scorned merely because one cannot hope to confront them with high-precision data. Well-constructed theories often can be challenged by semiquantitative experiments.

The next two chapters deal with changes in populations that survive but a single year or season. Chapter 2 is concerned with general population levels, and Chapter 3 deals with shifts in genotypes at a single locus. Both chapters deal with a probably unfamiliar yet simple and powerful type of mathematics: difference equations. These chapters also introduce for the first time the recurring themes of stability and qualitative dependence on parameters – themes that are illustrated again (but for differential equations) in Chapter 5 in a study of homeostasis in the chemostat.

Biochemical kinetics is an area in which mathematical modeling has played an indispensable role. Certain aspects of this subject are considered in Chapter 4, notably a careful derivation of the "quasi-steady-state assumption." This assumption is very widely used, but relatively few have a clear idea of its limitations. It is a salutary experience to think carefully about a commonly employed approximation, for a skepticism is engendered that is healthy (if it is not overemployed).

Chapter 6 uses many concepts that were gradually introduced earlier (quasi-steady state, dimensionless variables, phase plane, etc.) to show how well-known biochemical interactions can automatically result in the observed complex behaviors of the cyclic adenosine 3′,5′-monophosphate (cAMP) secretion system in cellular slime molds. It is not yet established whether or not the central idea of the mathematical model is correct for cAMP signaling in slime molds, but this and related theories have already had a noticeable impact on the field. In addition, the idea that switches in behavior accompany slow traverses of parameter space has now been suggested in several other important contexts (e.g., in the work on morphogenesis that is discussed toward the end of Chapter 9). Switches occur so generally in the type of equations which characterize biological systems that the ideas enunciated almost certainly will prove of considerable importance in biology.

So as not to stray too far from our principal goals, only in Chapter 6 do we give a fairly comprehensive survey of the experimental evidence in favor of the theory that has been presented.

Chapter 7 introduces partial differential equations, in the context of diffusion. For the first time, some knowledge of integral calculus is required (to understand the derivation of the diffusion equation). It is shown that one can easily *verify* the purported solution to a partial differential equation even though one has not had the training necessary to *deduce* the solution. A "rule of thumb" for estimating diffusive effects is derived and applied. Chapter 8 concentrates on the remarkable morphogenetic (pattern-forming) potentialities of a system of reacting and diffusing chemicals. Most readers will find a challenge in the mathematics, concerned as it is with stability theory for partial differential equations. Chapter 9 also deals with morphogenesis, but the emphasis here is on the role of mechanical forces. There is an instructive contrast with the approach of Chapter 8.

Chapters 2, 3, 5, 6, 8, and 9 provide examples, exhibiting gradually increasing difficulty, of a general approach to the study of dynamic problems – find the steady states, examine their stability, conjecture global qualitative behavior as a function of parameters, check your conjecture with selected computer simulations. Elucidation of this approach is the central mathematical theme of this book.

A word should be said about the computer. This book requires no knowledge of machine computation, although some exercises have been provided for the application of hand calculators, simple or programmable, and/or larger-scale machines. It would be better if the role of the computer could be made more central – and, indeed, at the Weizmann Institute the differential equations solver CSMP has been explained and utilized for simulations in some versions of the course. The limitations of what can realistically be taught in a single semester have led us to abandon direct instruction in computer techniques. Instead, we try repeatedly to illustrate the importance of machine computation, with the intention of motivating the reader to undertake further study of this subject. It is recommended that biologists learn an elementary language such as SPEAKEASY or BASIC, for remarkable results can be obtained after a few hours of training. Spending a little time learning CSMP (or a comparable high-level program "package") is also very worthwhile, as is illustrated by the fact that the following is a complete program for solving the nonlinear differential equation

$$ds/dt = 7s/(9+s), \quad 0 < t < 10, \quad s(0) = 15,$$

and tabulating and graphing the result:

```
F = 7. * S/(9. + S)
S = INTGRL(15.,F)
TIMER FINTIM = 10
PRINT S
OUTPUT S
END
STOP
```

Biologists do not need to spend years learning mathematical methods before they can apply theory to their work. With a study of concepts such as those presented here, plus as little as a few hours of training in computer programming, in many cases they are already prepared to improve significantly the quality of their research by providing it, when needed, with an adequate theoretical framework.

The computer revolution makes theoretical work an essential part of contemporary biological education for three reasons. (1) Computer-controlled instrumentation (and other technological innovations) can provide data of a quantity and quality that was only a dream a few years ago. (2) Theories are required to organize masses of existing data and to guide future data-acquisition efforts. (3) Software packages (such as CSMP) make it possible for relative novices to do highly respectable theoretical work with the aid of a computer – once they have had experience with some basic conceptual frameworks that are used by theorists. The provision of such experience is the main aim of this book.

The reader should be aware of the fact that a somewhat special approach must be used in reading theoretical texts. The discussion is filled with statements of the form "From ( ) and ( ) it follows that ( )," where the parentheses indicate certain equations or inequalities. On a first reading it is probably best to accept a statement of this type and read on, trying to catch the flow of the argument. There will come a point when the significance of various statements begins to be elusive. Then it is time to reread, making sure that the definitions of the various terms are beginning to be fixed in your mind. If you forget a definition, look it up again. Work actively, with a pencil and scratch paper. It is worthwhile on a second or third reading to verify some of the equations. This will help sharpen your technical skills and will further clarify and emphasize the definitions and concepts. When the details of a section begin to be clear, read it once more, and then go on.

Students usually find the material rather difficult at first. Understanding grows as the semester continues. Practice builds confidence, and major concepts become familiar through repeated application. Appreciation of the material also increases, because problems of greater biological importance can be tackled when better tools are at hand.

Seven classes of first-year graduate students in biology at the Weizmann Institute helped form this text – by their tacit disapproval or positive responses in class, and by explicit suggestions. The author extends his gratitude to all these men and women, and also to Volker Brendel, Arthur Mazer, and Daniel Segel for carefully reading through the whole text and pointing out errors. Albert Goldbeter, Hanna Parnas, Graeme Mitchison, and Hans Meinhardt kindly made suggestions on the material based on their work. Leah Edelstein-Keish and Frank Rothman made several helpful remarks.

Thanks are also due to various authors and publishers for allowing reproduction of figures. (The legends to the figures describe the precise sources of reproduced material.)

Doris Budowski, Michelle Bensimon, and Carol Weintraub expertly performed the typing. Special thanks are due to Yehudah Barbut for his skilled and careful drawing of the figures. The editorial staff of Cambridge University Press have been most helpful, particularly David Tranah. Finally, Gary Odell must be singled out for his kindness in providing phase-plane diagrams (and the graphics for the cover) with his own elegant phase portrait package and a Ramtek color graphics computer facility purchased by NSF grant MCS-77-27493.

Note to the instructor

More and more biologists are coming to agree that theoretical analyses can be of great value in certain aspects of their subject and that the theory becomes increasingly essential as experimental information grows more detailed and profound. If this view is correct, an introduction to theoretical biology should form part of a first-rate educational program in the biological sciences. Such an introduction is not easy to provide, however, for prospective students will respond only to theories that genuinely advance their understanding of biology, yet their analytical background is not strong. Much classroom experience is required before a good course emerges; such experience led to the preparation of this volume.

This book is based on a course taught by the author since 1975 to first-year graduate students in the biological sciences at the Weizmann Institute of Science. The course is now compulsory for all such students, for it has been deemed successful in its goal of introducing biologists to the power of theoretical analysis. It lasts one semester and consists of two lecture hours and one recitation hour per week. The author's lectures concentrate on mathematical modeling and its significance. The recitation is taught by an assistant, who reviews calculus and later on teaches some new mathematical techniques (which are almost immediately applied in the lectures).

The students who take the course come from undergraduate institutions throughout the world. Their backgrounds are extremely varied. Their mathematics training typically consists of a good one-year introductory course in calculus, in their first year of university. They generally have not used much mathematics in their subsequent undergraduate studies. Consequently, they have forgotten virtually all of what they had learned of calculus – to the extent that when asked on the first day of class to differentiate x^2 with respect to x, many cannot answer correctly.

The main purpose of the course, and of this book, is to demonstrate the usefulness of mathematical modeling in the biological sciences, chiefly in molecular and cellular biology. Examples are selected to illustrate several major biological areas in which theoretical work is important.

There is a gradual buildup of mathematical sophistication. In our experience, the material can be well handled with the aid of a calculus review (as outlined in Appendix 1) that stresses the meaning of the various elementary mathematical concepts and refreshes basic manipulative skills. Forgotten material is quickly recalled if it is concisely but clearly reviewed and then put to immediate use in a problem of interest to the student.

Some new mathematical topics are introduced, chiefly in connection with elementary difference and differential equations. Experience teaches how to strip this down to the bare bones, a necessary task if there is to be the requisite time to develop the applications with the care required. The appendixes document this experience.

Although no advanced mathematical manipulations are used, fairly advanced mathematical *concepts* are treated. A climax in this sense is reached in Chapter 6, which deals with the fact that for phenomena governed by nonlinear ordinary differential equations, different domains of parameter space may be associated with different qualitative behaviors, with bifurcations occurring as the state of the system passes from one domain to another. By building up gradually to this climax, one can present these theoretical ideas so that they can be thoroughly understood by biologists. Chapter 9 presents similar conceptual material in another context. The use of such nontrivial mathematical concepts in discussing important biological problems indeed demonstrates that mathematics can on occasion play a vital role in biological investigations.

All the material in this book has been successfully taught to biologists, most of it several times. Chapter 8 formed part of a more advanced course, but it was negotiated without undue difficulty by biology students whose background came from Chapters 1 through 7.

The book contains considerably more material than can comfortably be covered in a semester by students with the minimal background. The course is entirely suited to advanced undergraduates – indeed, it would be advantageous to integrate it into the undergraduate education of biologists. A one-year course could be nicely composed by combining Newby (1980), for example and the present text.

In principle, the text requires no previous knowledge of biology – but of course the importance of the topics treated can be accurately estimated only by persons with a good biological basis. Thus, although the book is primarily intended for biologists, it can also be used by mathematicians who seek important and interesting applications of basic analysis that do not require an extensive scientific background. The very first chapter, to give one example, treats an optimization problem of considerable importance in which the extremum is found at a point where the function in question is not differentiable. By contrast, Chapter 8 employs a number of relatively advanced topics – stability analysis of a system of partial differential equations, Fourier series, numerical analysis – while considering a central class of models in developmental biology. The main theoretical line – analysis of bifurcations and their consequences – is a primary theme of applied mathematics.

If the book is to be used in a course for biologists, the more difficult mathematical portions probably ought to be omitted. For example, there is no loss in continuity in treating only the simplified model in Chapter 1 and skipping the latter part of Chapter 2 (dealing with period doubling and chaos). If most of the rest of the text through Chapter 6 is thoroughly covered in a semester, the students will have been exposed to a great deal. If a little more time is available, even a brief overview of Chapter 9 (which requires no new technical ideas) will provide strong reinforcement of much novel material in Chapter 6.

Chapter 7 is elementary, but it requires a review of integration. Chapter 8 will be found relatively difficult by the mathematically unsophisticated, although it builds only on what has preceded it. The material up to "Patterns in finite intervals" presents the essence of the matter and is entirely in keeping with the gradually increasing level of sophistication in the earlier part of the book.

We conclude with an outline of the organization of the book. It will be seen that there is considerable flexibility in selecting and arranging the material to be presented.

Appendix 1 (Mathematical prerequisites) is required throughout. Additional prerequisites for the various chapters are as follows:

Chapter	Further required material
1	None
2	Appendixes 2 and 3
3	Appendixes 2 and 3
4	None
5	Appendix 5
6	Appendix 5
7	Appendix 8
8	Chapter 7, Appendix 5
9	Appendix 5

Appendixes 2 and 4 are prerequisites to Appendix 5. Strictly speaking, Appendix 6 (Complex numbers) is needed to handle oscillatory solutions of the second-order differential equations and systems treated in

Appendix 4. Nonetheless, study of Appendix 6 can be deferred by verifying (using direct substitution) the transition from equations (A4.14) to (A4.17), or by temporarily taking this transition on faith.

Dimensionless variables are employed many times, in a self-contained manner. The overall advantages of this practice can be treated at any point in the course by presenting Appendix 7.

Conventions and notations

Equations are numbered consecutively within each chapter. If equation (3) in Chapter 4 is referred to in another chapter, it is referred to as equation (4.3). The second figure in Chapter 6 is always referred to as Figure 6.2.

The appendixes at the end of the book provide mathematical material. Some chapters have supplements in which certain additional material is presented. Equation (A3.2) is the second equation of Appendix 3. Equation (S7.3) is the third equation in the supplement to Chapter 7. Exercise 1.3 is the third exercise in Chapter 1. Hints or answers to exercises marked by a dagger (†) are given after the appendixes.

The notation "$\equiv$" means "is identically equal to," as in $\sin^2 x + \cos^2 x \equiv 1$. It is also used in definitions. For example, if $x(t)$ denotes position at time t, then velocity v is defined by $v \equiv dx/dt$.

The notation

$$\frac{df(x)}{dx} \bigg|_{x=t^2}$$

means that one should first determine $df(x)/dx$ and then make the substitution $x = t^2$. A similar-appearing notation, used in connection with integration, is

$$F(x) \bigg|_a^b \equiv F(b) - F(a).$$

References, given the form "Smith (1981)," are collected at the end of the text. Important equations are indicated by double diamonds ($\blacklozenge\blacklozenge$).

1

Optimal strategies for the metabolism of storage materials

This chapter demonstrates that elementary calculus can be used to illuminate some aspects of the general question of how organisms should employ storage materials, given that they have evolved to be "optimum" in some sense. The material presented here is a slight generalization and expansion of ideas discussed by Cohen and Parnas (1976).

Biological background

Storage materials enable organisms to survive and prosper when there is a temporary shortage of energy. We shall consider only microorganisms in our discussion. For these, a commonly held view is the "excess theory," according to which microorganisms synthesize storage materials if and only if the supply of external nutrient exceeds the supply necessary for maximal current growth. The approach to be taken here is different and is based on the following hypothesis.

Assumption 1. Storage materials are synthesized in such a manner as to maximize the long-term growth rate.

We shall study photosynthetic algae under circumstances in which a day of uniform light intensity and fixed length is followed by a night with no light. The algae can utilize the light and available nutrients to synthesize proteins required for growth and reproduction. During the day they can also synthesize energy-rich storage materials. These can be used to furnish some or all of the energy that must be supplied at night, when there is no energy flow from the environment and no synthesis – but when energy is required for maintenance and possible cell division. The starchy storage materials supply energy efficiently, but their presence has a cost in diminishing protein production. The sort of question that we shall address here is How much storage material should be synthesized by optimally behaving algae under various possible circumstances? The reader might pause now to try to answer this question, to see whether or not unaided intuition can arrive at the same conclusions that will be generated by our theoretical arguments.

There is evidence that photosynthesis is depressed during the time in which cells divide. We shall assume that the depression is sufficiently large that division during the night will maximize the expected number of daughter cells.

Assumption 2. Cell division takes place only at night.

At any time t we shall regard a cell as being composed of just two types of materials: biosynthetic materials, denoted by $P(t)$, and storage materials, denoted by $S(t)$. The synthetic materials are largely proteins. In the present context we thus regard "protein" and "synthetic material" as synonymous.

Consider the algae that are contained in some experimental setup, or in some defined natural region. At the beginning of the day, $t=0$, let P_0 and S_0 denote the amounts of protein and storage material that are associated with these algae.[1] [Thus, $P(0)=P_0, S(0)=S_0$.]

Assumption 3. All cellular materials are synthesized at a constant *photosynthesis rate R* per unit amount of synthetic material.

At a given moment, both protein and storage material are synthesized at rates that are proportional to the amount of synthetic material (i.e., protein) that is present. Consequently, as we shall see in detail later, the amount of protein increases exponentially throughout the time it is being synthesized. Suppose that the cell's "best strategy" requires it to synthesize a certain amount of storage material. The amount of protein at the end of the day will be largest, as required by Assumption 1, if the least possible time is devoted to synthesis of the required storage material. This is accomplished by postponing the period of storage synthesis to the end of the day, when the amount of synthetic machinery is maximal.

The postponement of storage synthesis is really a conclusion of our analysis. We have derived it intuitively, but a rigorous derivation could be given with the aid of optimal control theory.

Conclusion 1. Storage material is synthesized at the end of the day.

1 A list of symbols and their definitions is provided in the supplement to this chapter. For later chapters, and in general when reading theoretical material, it is recommended that the reader compile such a list.

Calculation of P and S at the end of the day

Given that its synthesis occurs at the end of the day, the amount of storage material is completely determined by the number of hours devoted to its production. Call this number τ (tau). *The purpose of our model is to calculate the optimal value of τ.* The first step is to find mathematical expressions for the amounts of P and S at the end of the day, for any given value of τ.

Let the length of the day be T hours (so that, of course, the length of the night is $24 - T$ hours). Because τ hours at the end of the day are devoted to the synthesis of storage material S, protein P will reach its maximum after $T - \tau$ hours and will remain the same for the rest of the day:

$$P(T) = P(T - \tau).$$

By Assumption 3, if the day is T hours long, it then follows that

$$S(T) = R\tau P(T). \tag{1}$$

[For full accuracy, the initial amount of storage material S_0 should be added to the right side of equation (1), but the relative contribution of this term is negligible.]

We here introduce an abbreviation that we shall employ frequently:

$$\sigma \equiv S(T)/P(T), \quad 0 < \sigma < \infty. \tag{2}$$

The importance of σ (sigma) is evident even at this early stage; it is the ratio of storage material to synthetic material that the cells produce. In economic terms it is the ratio of "capital" (invested for future needs) to production equipment. In terms of (2), equation (1) becomes

$$\tau = \sigma/R. \tag{3}$$

Because R is fixed, instead of calculating the optimal value of τ we can calculate the optimal value of σ – and, indeed, this latter calculation is slightly more convenient.

Assumption 3, that synthesis takes place at a constant rate R, implies that at any time t, $0 < t \leqslant T - \tau$,

$$dP(t)/dt = RP(t).$$

From the fact that P_0 is the amount of P present at the beginning of the day, it follows (Exercise A1.7) that

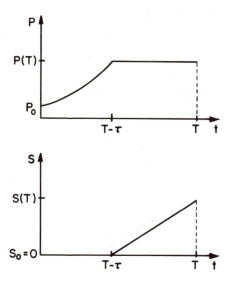

Figure 1.1. Graphs illustrating the result that exponential accumulation of synthetic material P for $T-\tau$ hours is succeeded by uniform synthesis of storage material S for τ hours. (T is the length of the day.)

$$P(t) = P_0 \exp(Rt). \qquad (4a)$$

Given that τ is the time allotted for the synthesis of storage material and T is the length of the day, P will be synthesized for $T-\tau$ hours, after which the amount of P will remain constant for the rest of the day. Thus, as we have seen, $P(T) = P(T-\tau)$. Consequently from (4a),

$$P(T) = P_0 \exp[R(T-\tau)]. \qquad (4b)$$

See Figure 1.1 for a graphical view of the accumulation of protein and storage material.

We note for later use that the number of new daughter cells N can be calculated from the formula

$$N = [P(T) - P_0]/W, \qquad (5)$$

where W is the average weight of a daughter cell.

Calculating the net "profit" from a given amount of storage material

We are examining how storage material enables organisms to operate successfully in the presence of an energy shortage – at night in the present instance. Storage material is a rich source of the required energy, but it is accumulated at the price of forgoing protein synthesis. We must now calculate precisely the net advantage that flows from the possession of a given amount of storage material – in the two cases in which this amount does or does not furnish the total nocturnal energy requirements.

Assumption 4. During the night, cells need E_M units of maintenance energy per unit weight of synthetic material per hour. Also, the formation of each daughter requires an amount of energy E_d. Thus, a total amount of energy E is required each night, where

$$E = E_M(24-T)P(T) + E_d N. \tag{6}$$

There are two possibilities:

     Case I: Stored energy is less than E.

     Case II: Stored energy is greater than E.

In Case II, the cell will convert the excess stored energy into synthetic material. In Case I, synthetic material must be broken down to provide the required energy. We symbolize the various "costs" by

$$P \xrightarrow{k_P} \text{energy}, \quad S \xrightarrow{k_S} \text{energy}, \quad S \xrightarrow{\beta} P. \tag{7}$$

That is, there are k_P (respectively, k_S) energy units per unit amount of P (respectively, S), and conversion of a unit of storage material yields β units of synthetic material.

In the two cases, I and II, we shall now form the expressions for the *gain* in nocturnal cell production due to the presence of storage material (which is an efficient source of required energy) and the *loss* due to the fact that synthesis of protein during the day had to be forsaken because of the production of storage material. We shall then seek a policy that will maximize the "profit," the gain minus the loss.

Note from (5) that only protein contributes to the formation of new organisms. Thus, we shall measure the *loss* in cell production by the total amount of protein L that was not synthesized during the *day*, because part of the day was devoted to storage synthesis. The *gain* is the total amount of protein G that did not have to be broken down during the *night*, because part of the energy requirements could be met by storage. Gain and loss are each figured out for a given policy. For each policy, one subtracts what did happen from the extreme case in which there is no synthesis whatever of storage material. In the computation of loss, for example, the relevant extreme case is that in which protein is synthesized all day. From this, the protein actually synthesized is subtracted.

Case I occurs when

$$k_S S(T) < E. \tag{8}$$

In this case the energy available from the stored material is not sufficient, so that an extra amount of energy must be supplied by breakdown of P. If x denotes the amount of protein that must be broken down, then

$$k_P x = E - k_S S(T), \quad \text{i.e.,} \quad x = [E - k_S S(T)]/k_P. \tag{9a,b}$$

If there were no storage material, all the necessary energy would have to be extracted from P. The amount of P required would be E/k_P. Subtracting from this the protein that is actually required in Case I, we have a gain G due to the presence of the storage material, given by

$$G = S(T)/\alpha, \quad \text{where } \alpha \equiv k_P/k_S. \tag{10a,b}$$

Case II occurs when $k_S S(T) > E$. Here, instead of having to break down protein to supply the nocturnal energy requirements, there is an excess of stored energy that can be converted into additional protein. Consequently, as the reader is asked to verify,[2]

$$G = \frac{E}{k_P} + \beta \frac{k_S S(T) - E}{k_S}. \tag{11}$$

In both Case I and Case II the loss L is the same: the amount of P that would have been synthesized if the whole day of length T had been utilized, minus the actual amount synthesized in $T - \tau$ hours. Thus, from (4a),

$$L = P_0 \exp(RT) - P_0 \exp[R(T - \tau)]. \tag{12}$$

Combining (10), (11), and (12), we can calculate the profit $\pi = G - L$, where by profit we mean the increase in synthetic material brought about by the storage policy. After a little rearrangement, we obtain (Exercise 1) the following formulas:[2]

◆◆   I: $\pi = P_0 \exp(RT)\left[\left(1 + \dfrac{\sigma}{\alpha}\right)e^{-\sigma} - 1\right]$ when $k_S S(T) < E$. $\qquad$ (13a)

◆◆   II: $\pi = P_0 \exp(RT)[(1 + \beta\sigma)e^{-\sigma} - 1]$

$$+ E\left(\frac{1}{k_P} - \frac{\beta}{k_S}\right) \quad \text{when } k_S S(T) > E. \tag{13b}$$

---

2 On first reading, (11) and (13) should be accepted. Verification should be attempted later, as explained in the Preface.

The final mathematical problem: maximize profit

By Assumption 1, we wish to maximize π. More precisely, because the possible storage "strategy" of the cells is expressed in their choice of σ, we wish to *select the synthetic/storage ratio σ that will maximize $\pi(\sigma)$*. The italicized phrase constitutes the mathematical problem to which we have been led by our model building.

A simplified model

We shall now make an assumption that will make the problem considerably simpler. To this end, note from (13) that Case II occurs if and only if $k_S S(T) > E$. Because σ is the variable with which we are concerned, it is natural to use (2) and to write the condition for Case II in the form

$$k_S \sigma > E/P(T). \tag{14}$$

But this key condition for distinguishing Cases I and II is not a simple one, because $E/P(T)$ depends on σ. To see this, first observe that (6) and (5) imply[3]

$$\frac{E}{P(T)} = E_M(24-T) + \frac{E_d N}{P(T)} = E_M(24-T) + \frac{E_d}{W}\left(1 - \frac{P_0}{P(T)}\right). \tag{15}$$

By (4) and (3), the right side of (15) depends on σ [through $P(T)$]. Consequently, (14), together with (15), presents a somewhat complicated inequality to be solved for σ.

We can circumvent the difficult nature of (14) by limiting our considerations to situations in which there is so much protein production (because of intense light) that

$$P_0 \ll P(T). \tag{16}$$

Now the "nonconstant" term $P_0/P(T)$ in (15) can be neglected.

To take advantage of our finding that $E/P(T)$ can be regarded as a constant if (16) holds, we reformulate our problem with the aid of the definition

$$\bar{\sigma} \equiv E/[k_S P(T)]. \tag{17}$$

We now see from (14) that Case II holds when $\sigma > \bar{\sigma}$. From (15) and (16) it follows that

---

3 Formula (15) can at first be accepted without checking, but a reader with pencil in hand (see Preface) should experience no difficulty in rapidly verifying it.

$$\bar{\sigma} \approx E_M(24-T)/k_S + (E_d/Wk_S) = \text{constant}. \tag{18}$$

Using (17) to replace E, we find that expression (13b) for the profit in Case II becomes

$$\pi = P_0 \exp(RT)\left\{e^{-\sigma}\left[\bar{\sigma}\left(\frac{1}{\alpha}-\beta\right)+\beta\sigma+1\right]-1\right\}. \tag{19}$$

Now our mathematical problem is to maximize the function $\pi(\sigma)$ given by (13a) when $\sigma < \bar{\sigma}$ and by (19) when $\sigma > \bar{\sigma}$, where $\bar{\sigma}$ is a constant.

Solution of the simplified mathematical problem

Let us begin with Case II, where $\sigma > \bar{\sigma}$. From (19) we can easily verify (Exercise 1) that

$$\partial\pi/\partial\sigma = P_0 \exp(RT-\sigma)[\beta - \beta\sigma - \bar{\sigma}(\alpha^{-1}-\beta)-1]. \tag{20}$$

We have a zero derivative, the classic necessary condition for a maximum, if

$$\beta\sigma = \beta - \bar{\sigma}\left(\frac{1}{\alpha}-\beta\right)-1, \quad \text{i.e., if } \sigma = \bar{\sigma} - \frac{\bar{\sigma}}{\alpha\beta} - \frac{1}{\beta}+1. \tag{21a,b}$$

But $\sigma > \bar{\sigma}$ in the region under consideration, so that σ is certainly greater than the right side of (21b) – for we expect that

$$\beta < 1, \quad \text{so that } \frac{1}{\beta} > 1, \quad -\frac{1}{\beta}+1<0. \tag{22a,b,c}$$

[Equation (22a) records the biological assumption that something is lost in converting storage material to protein; see the definition of β in (7).] We thus see that the derivative in (20) is always negative, and the largest value of $\pi(\sigma)$ in Case II occurs on the boundary of the region, where $\sigma = \bar{\sigma}$. Let us now check Case I. Here we find, from (13a) [Exercise 1], that

$$\partial\pi/\partial\sigma = P_0 \exp(RT-\sigma)(1-\sigma-\alpha)/\alpha. \tag{23}$$

The derivative vanishes when

$$\sigma = 1-\alpha. \tag{24}$$

Note that the right side of (24) will be positive, because $\alpha \equiv k_P/k_S$ is certainly expected (Exercise 2) to satisfy

$$\alpha < 1. \tag{25}$$

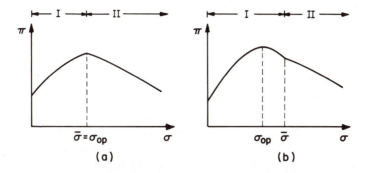

Figure 1.2. Schematic shape of the profit function (in the simplified problem) when (a) $1-\alpha>\bar{\sigma}$, (b) $1-\alpha<\bar{\sigma}$. Note that in case (a) the derivative does not vanish at the maximum; the derivative at $\bar{\sigma}$ does not exist.

The point $\sigma=1-\alpha$ is within the region $\sigma<\bar{\sigma}$ if and only if

$$1-\alpha<\bar{\sigma}. \tag{26}$$

If (26) does not hold, the derivative when $\sigma<\bar{\sigma}$ is positive, and the maximum value of π within this region occurs when $\sigma=\bar{\sigma}$. See Figure 1.2.

Let σ_{op} be the optimal value of the ratio $\sigma=S(T)/P(T)$, that is, the value that gives a maximum profit $\pi(\sigma)$. Summarizing the results of our calculations we obtain the following.

Conclusion 2

$$\sigma_{op}=1-\alpha \quad \text{if } 1-\alpha\leqslant\bar{\sigma}, \tag{27a}$$

$$\sigma_{op}=\bar{\sigma} \quad \text{if } 1-\alpha\geqslant\bar{\sigma}. \tag{27b}$$

In practice, typical values of the parameters are $0.1\leqslant\bar{\sigma}\leqslant0.25$ and $\alpha=0.4$ (Parnas and Cohen, 1976, p. 36). Thus, (27b) holds, and we predict that *storage materials are generally produced during the day in the exact amount needed at night.*

Conclusions from the mathematical results

We shall be content to draw only one further conclusion from the present results. This requires studying the effect of the rate of photosynthesis R on the optimal time interval τ_{op} during which storage material is synthesized. Assuming that the algae have evolved to operate under optimal conditions, we note that (27a) and the relation $\tau_{op}=\sigma_{op}/R$ of (3) yield

$$\partial\tau_{op}/\partial R=-(1-\alpha)/R^2<0. \tag{28}$$

If (27b) holds, then, similarly,

$$\partial \tau_{op}/\partial R = -\bar{\sigma}/R^2 < 0. \tag{29}$$

Conclusion 3. The optimal length of time for the synthesis of storage material decreases when photosynthesis increases.

Conclusion 3 is particularly interesting, for it is opposite to what is expected from the excess theory. In fact, Assumption 1 and Conclusion 1 have been verified for a number of microorganisms. Conclusion 3 has been verified by Cohen and Parnas (1976) for *Chlamydomonas reinhardii.*

The full problem

We have proceeded under the assumption that the last term in (15) can be neglected. When light intensities are strong, as many as eight daughter algae appear during the night from the division of a single large cell; in this case the simplification is justified. On the other hand, when light intensities are weak, we can still proceed – without simplifying – by employing a little more sophisticated mathematics.

We recall that, in general, $\bar{\sigma}$ is not constant (i.e., $\bar{\sigma}$ depends on σ); this is the root of our difficulty. To proceed, it is convenient to define the constants ψ (psi) and ϕ (phi) by

$$\psi \equiv \frac{E_M(24-T)}{k_S} + \frac{E_d}{k_S W}, \quad \phi \equiv \frac{E_d}{k_S W} \exp(-RT). \tag{30a,b}$$

With these constants, the exact expression for $\bar{\sigma}$ is

$$\bar{\sigma}(\sigma) = \psi - \phi e^{\sigma}, \tag{31}$$

combining (17), (15), and (4b). Consequently,

$$\sigma > \bar{\sigma} \quad \text{if and only if} \quad \sigma > \psi - \phi e^{\sigma}. \tag{32a,b}$$

For (32b) to hold, the graph of $y = \sigma$ must lie above the graph of $y = \psi - \phi e^{\sigma}$. These graphs are depicted in Figure 1.3. It follows from the figure that

$$\sigma > \bar{\sigma} \quad \text{if and only if} \quad \sigma > \sigma^*. \tag{33}$$

Here σ^* *is* a constant, defined by the intersection of the two graphs:

$$\sigma^* = \psi - \phi \exp(\sigma^*). \tag{34}$$

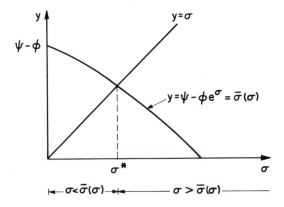

Figure 1.3. Graph for determining when σ is less than or greater than $\bar{\sigma}(\sigma) \equiv \psi - \phi e^{\sigma}$.

Our problem is now seen to require the maximization of a function $\pi(\sigma)$ given by (13a) when $\sigma < \sigma^*$ and by (19) when $\sigma > \sigma^*$.

Let us first consider Case II, when $\sigma > \sigma^*$. Noting from (31) that

$$\frac{\partial \bar{\sigma}}{\partial \sigma} = -\phi e^{\sigma}, \tag{35}$$

we differentiate (19) to find (Exercise 3) that for Case II,

$$\frac{\partial \pi}{\partial \sigma} = P_0 \exp(RT - \sigma)\left[-(1-\beta) - \beta\sigma - \left(\frac{1}{\alpha} - \beta\right)(\bar{\sigma} + \phi e^{\sigma})\right]. \tag{36}$$

The positivity of $\bar{\sigma} + \phi \exp(\sigma)$ follows from (31). Consideration of (22a) and (25) thus shows that $\partial\pi/\partial\sigma$ remains negative when $\sigma > \bar{\sigma}$ even when the simplifying assumption (16) is not made.

Because $\bar{\sigma}$ does not appear in (13a), $\partial\pi/\partial\sigma$ is computed in region I as before. We conclude that if simplification (16) is not made, Conclusion 2 [equation (27)] must be modified to

$$\sigma_{\mathrm{op}} = 1 - \alpha \quad \text{if } 1 - \alpha \leqslant \sigma^*, \tag{37a}$$

$$\sigma_{\mathrm{op}} = \sigma^* \quad \text{if } 1 - \alpha \geqslant \sigma^*. \tag{37b}$$

We have no explicit expression for the constant σ^*. But from (34), by implicit differentiation (Exercise 3), we can find the useful formula

$$\frac{\partial \sigma^*}{\partial R} = \frac{T\phi}{\phi + e^{-\sigma^*}}. \tag{38}$$

Equation (38) shows that when $1 - \alpha \geqslant \sigma^*$, σ_{op} is an increasing function of the photosynthesis rate R. Equation (38) is also necessary in recalculating (29), for when $1 - \alpha \geqslant \sigma^*$,

$$\tau_{op} \equiv \frac{\sigma^*}{R}, \quad \frac{\partial \tau_{op}}{\partial R} = \frac{R(\partial \sigma^*/\partial R) - \sigma^*}{R^2}. \qquad (39a,b)$$

It might be argued that it is "obvious" that the optimal length of time τ_{op} during which storage material should be photosynthesized must decrease as the intensity of sunlight increases (Conclusion 3). But in (39b) our more general model shows that the sign of $\partial \tau_{op}/\partial R$ can, in principle, be either positive or negative – so that the behavior of τ_{op} could be counterintuitive. In fact, on substituting typical numerical values, it can be shown that $\partial \tau/\partial R$ retains the negative sign found in (29).

Final remarks

This chapter can be regarded as giving one example of an optimization approach that has been used in many biological contexts. Another example is provided by optimum foraging studies that proceed under the hypothesis that animals adopt feeding strategies that maximize the rate of net energy gain. See Hainsworth and Wolf (1979) for a review and Townsend and Calow (1981) for a recent example of several texts concerned with the adaptive significance of physiological characteristics.

A particularly extensive and interesting treatment of optimization arguments in biology can be found in the Oster and Wilson (1978) study of social insects. Of special value is the critique of optimization theory found in the last chapter. For example, these authors point out that optimization arguments, as in the present case, leave open the question of how the optimum strategy is implemented. (This is not a fatal flaw, because successful theories may well raise more interesting questions than they answer.) They remark that careful examination of a biological situation often reveals several conflicting goals, so that the question may not be one of optimization, but rather how "conflicting interests...can be resolved in a stable fashion." They stress the crucial difference between engineers who use optimization theory to arrive at a most efficient design and theoretical biologists who use the theory "to infer 'nature's design' already created by natural selection." It might also be added that engineers can, in principle, build structures in an entirely novel fashion, whereas organisms can improve themselves only relatively gradually, by tinkering with their existing structure. In addition, some traits may not be adaptive at all, as evolution lags a changing environment or promotes some traits as side effects of others (Maynard Smith, 1978).

In short, optimization theory has already provided many biological insights, but, perhaps even more than other theoretical frameworks, it must be employed with taste and restraint.

Exercises

1. Verify equations (11), (13), (20), and (23).
2. Why is (25) expected on biological grounds?
3. Verify equations (36) and (38).
4. How would the various formulas and conclusions be changed if Assumption 3 were modified to take into account different constant rates of synthesis for P and S?
5. Try to supply convincing intuitive reasons to support result (27a) that under certain circumstances the optimal cell behavior is to synthesize (during the day) less storage material than is required to provide the energy needed at night.
6. (a) Show that the function $\pi(\sigma)$ is continuous at $\sigma = \bar{\sigma}$ by demonstrating that $\pi(\sigma)$ has the same limit as $\sigma \to \bar{\sigma}$, whether $\sigma < \bar{\sigma}$ or $\sigma > \bar{\sigma}$.
   (b) Show that different answers are obtained for

   $$\lim_{\sigma \to \bar{\sigma}} \frac{d\pi(\sigma)}{d\sigma}$$

   depending on whether $\sigma < \bar{\sigma}$ or $\sigma > \bar{\sigma}$ when the limit is taken. That is, show that $\pi(S)$ has a discontinuous derivative at $\sigma = \bar{\sigma}$.
   (c) Show that $\psi - \phi > 0$, as is assumed in Figure 1.3. In addition, show that $\bar{\sigma}(\sigma) \geq 0$ for all biologically reasonable parameter values.
7. Implicit in equations (5) and (6) is the assumption that not much of the synthetic material $P(T)$ is broken down at night to supply energy. Conclusion 2 indicates that this assumption is generally valid. Nonetheless, even though cases in which $P(T)$ is significantly broken down occur only rarely, discuss the extensions of the mathematical model that would be required to deal with this phenomenon. [Remarks: (i) The problem is not fully specified, as is of course true for most research problems. For example, one needs to know something about the timing of the nocturnal cell divisions. (ii) A full discussion would assume the dimensions of a miniproject. Even with very limited time, however, something can be learned by beginning an account of some of the issues involved and by making some explicit, not-unreasonable assumptions concerning the biology – which properly should be the subject of literature review and/or experiment.]
8. What do you anticipate to be the behavior of the fraction of storage materials as a function of light intensity for algae cells growing in continuous light? After pondering the matter, compare your thoughts with those expressed by Parnas and Cohen (1976, pp. 10–11).
9. By checking the second derivative, verify that (24) provides a minimum [when (25) holds].

† 10. (a) Consider the function f given by

   $$f(x) = x^2 \quad \text{when } x \leq 1;$$
   $$f(x) = [(x-b)/(1-b)]^2 \quad \text{when } x \geq 1, \ b \neq 1.$$

For what value of x is f a minimum? A maximum? Sketch the graph of f.

(b) Repeat (a) for the function given by

$$f(x) = x^2, \quad -\infty < x \leqslant 1;$$
$$f(x) = 4 - 3x, \quad 1 \leqslant x.$$

11. Sketch the graph of $f(x)$ for $0 \leqslant x \leqslant 4\pi$ if

$$f(x) = x \quad \text{when } \sin x \geqslant \cos x;$$
$$f(x) = -x \quad \text{when } \sin x < \cos x.$$

12. Consider the function f given for $x \geqslant 1$ by

$$f(x) = x^2 - 4x \quad \text{when } \ln x - x + 2 \geqslant 0;$$
$$f(x) = (\ln x + 2)^2 - 4x \quad \text{when } \ln x - x + 2 \leqslant 0.$$

(a) By sketching on one graph the curves $y = \ln x$ and $y = x - 2$, show that the equation $\ln x - x + 2 = 0$, $x \geqslant 1$, has a single solution x^*, where $x^* > 2$.

(b) Show that the function f is continuous at $x = x^*$.

†

(c) Show that if $g(x) = (\ln x + 2)^2 - 4x$, then $dg/dx \leqslant 0$ for $x \geqslant 1$.

(d) Sketch the graph of $f(x)$ for $x \geqslant 1$. In particular, show that there is a maximum at $x = x^*$.

(e) Sketch the graph of $f(x)$ for $x \geqslant 1$ if

$$f(x) = (\ln x + 2)^2 - 4x \quad \text{when } \ln x - x + 2 \geqslant 0;$$
$$f(x) = x^2 - 4x \quad \text{when } \ln x - x + 2 \leqslant 0.$$

The following is taken from a discussion by A. Perelson (Segel, 1980, Section 5.3) of histamine release in basophil cells. Virtually no knowledge of the biology is required to answer the various questions.

13. Let C, m, and M be concentrations of free, singly bound, and doubly bound antigen molecules. At steady state, it is found that

$$M = \frac{mK_i(S_0 - m)}{2(1 + mK_i)}, \tag{40a}$$

where m satisfies

$$(1 - \beta)K_i m^2 + m - \beta S_0 = 0; \tag{40b}$$
$$\beta \equiv KC/(1 + KC).$$

Here K and K_i are dissociation constants, and S_0 is the total receptor concentration.

(a) Show that there is exactly one meaningful solution of the quadratic equation (40b).

(b) Show that on substituting the solution of (40a) into (40b) we obtain

$$M = S_0 \frac{1 + 2\delta - (1 + 4\delta)^{1/2}}{4\delta}, \quad \delta \equiv \beta(1 - \beta)K_i S_0.$$

Find the constants a_0, a_1, and a_2 in the series

$$(1 + 4\delta)^{1/2} = a_0 + a_1\delta + a_2\delta^2 + \dots,$$

and use the result to derive the approximation

$M \approx \frac{1}{2} S_0 \delta, \quad 0 < \delta \ll 1.$

(c) It is desired to find the value of β that maximizes M. To do this, it is sufficient to set $dM/dm = 0$ (m, *not* β!). Why? Show from (40a) that $dM/dm = 0$ if $m^2 K_i + 2m - S_0 = 0$.

(d) Conclude that M is maximized when $\beta = \frac{1}{2}$.

14. (a) Calculate the net amount of protein synthesis, over the day and night, if no storage material is made.

(b) Repeat (a) when τ hours at the end of the day are used to synthesize storage material.

(c) Use (a) and (b) to provide an alternative derivation of (13).

Supplement: List of symbols and their definitions

$P(t)$	Weight of protein at time t
$S(t)$	Weight of storage material at time t
P_0	Weight of protein at beginning of day
S_0	Weight of storage material at beginning of day
R	Rate of synthesis of cellular material per unit amount of P
τ (tau)	Hours per day devoted to production of S
T	Length of day
σ (sigma)	$S(T)/P(T)$
W	Average weight of daughter cell
E_M	Maintenance energy/unit weight of P/hour
E_d	Energy required for the formation of each new cell
E	Total energy required each night, for maintenance and reproduction
k_P	Energy units/unit weight of P
k_S	Energy units/unit weight of S
β (beta)	Units of S obtained on conversion of one unit of P
α (alpha)	k_P/k_S
G	Gain in P, owing to storage
L	Loss in P, owing to storage
π	Profit: gain minus loss
$\bar{\sigma}$	$E/[k_S P(T)]$
σ_{op}	Value of σ that maximizes profit
ψ, ϕ (psi, phi)	Combinations of constants defined in (33)
σ^*	Particular value of σ depicted in Figure 1.3
τ_{op}	Value of τ, when $\sigma = \sigma_{op}$

2

Recursion relations in ecological and cellular population dynamics

To focus ideas, consider a certain species of insect that emerges from its egg stage in the spring, lays eggs in the fall, and dies in the winter. Let N_i be the number of female insects within some fixed area, in the ith year, and assume that the number of males equals the number of females. A reasonable first guess is that the insect population next year will depend on this year's population level; that is,

$$N_{i+1} = f(N_i) \tag{1}$$

for some function f.

Given an initial population level N_0, solutions to (1) can be obtained recursively, one after another: $N_1 = f(N_0)$, $N_2 = f(N_1)$, and so forth. Thus, equation (1) is sometimes called a **recursion relation**. Because differences between populations from year to year can be calculated, (1) is also called a **difference equation**. Basic mathematical tools for the study of such equations are discussed in Appendix 3.

In the initial and major part of this chapter, relatively simple special cases of (1) will be discussed in the context of population dynamics. A number of important theoretical ideas will be presented for the first time, such as steady states, stability, and change in qualitative behavior as a function of parameter value. In a more advanced section we shall discuss the fact that even simple cases of (1) can exhibit extremely interesting and complicated behavior that sometimes seems indistinguishable from randomness. An application to cellular physiology will be presented.

Constant birthrate

The simplest case of equation (1) occurs when $f(N_i) = RN_i$, with R a constant. Here,

$$N_{i+1} = RN_i. \tag{2}$$

The constant R is seen to represent the net birthrate (birthrate minus death rate). As shown in Appendix 3, equation (2) can easily be solved explicitly. The behavior of the solution is described in Table A3.1. The major findings are in no way surprising. If the birthrate R is less than unity, N_i tends to zero: The insects will become extinct if each generation fails to reproduce itself. On the other hand, if each generation more than reproduces itself, the population (according to the constant-birthrate model) will grow without bound. In mathematical language,

$$R > 1 \quad \text{implies} \quad \lim_{i \to \infty} N_i = \infty. \tag{3}$$

Birthrate that decreases linearly with population level

A prediction that a population will possess a constant birthrate and hence will grow indefinitely is obviously incorrect. At higher population levels, various factors, such as depletion of resources and accumulation of toxins, will lead to a decrease in the birthrate. To model this decrease in the simplest possible way, let us suppose that the birthrate R decreases linearly as the population level N_i increases:

$$R = R(N_i) = r[1 - K^{-1}N_i], \quad r \text{ and } K \text{ positive constants.} \tag{4}$$

Thus, from (2),

$$N_{i+1} = rN_i[1 - K^{-1}N_i]. \tag{5}$$

We have written the constant preceding N_i in (4) as K^{-1}, for then K can be given biological meaning: When $N_i = K$, the population level is so high that there are no births whatever. When N_i exceeds K, the birthrate is negative, according to (4). [The reader should draw a graph of $R(N_i)$, as given in equation (4).] A negative birthrate is meaningless biologically, so that we can draw conclusions from our present model only when N_i remains less than K [which usually turns out to be so for the cases we consider – see Exercise 1(b)].

A change of variable frequently renders a problem simpler in form. Knowing this, it is quite natural to introduce the new variable x_i, defined by

$$x_i = N_i / K. \tag{6}$$

Thus, x_i provides the ratio of the population level in the ith generation to the borderline level K. With (6), equation (5) becomes

$$x_{i+1} = rx_i(1 - x_i). \tag{7}$$

In contrast to (5), which contains the two parameters r and K, (7) contains only the single parameter r. This decrease in parameter number is the benefit obtained by casting the problem in terms of the population *ratio* x_i. (See Appendix 7.)

We shall now study the behavior of solutions to (7). Before proceeding, the reader is urged to predict the outcome of our mathematical analysis. For example, does the population level always reach a steady value? If not, does it do so for some values of x_0? For some values of r? Which values? Under what conditions does the population become extinct? Can the population grow without bound? Does the population level ever oscillate back and forth (two-year cycles)? Are there ever three-year cycles? Four-year cycles? Five-year cycles? Remember, the predictions concern only the behavior of the highly simplified mathematical model (7). Effects on larval–pupal transitions of inevitable droughts, for example, though perhaps truly important, can play no legitimate role in a discussion of (7). Of course, such effects could influence the formulation of a more accurate model.

It is doubtful that the reader will accurately predict the behavior inherent in (7), for the full complexity of this behavior is only now becoming apparent even to research mathematicians. This illustrates one of the main reasons for constructing and analyzing mathematical models: Unaided intuition frequently is unable to fathom all the consequences of even fairly simple interactions.

Steady solutions to equation (7) and their stability

Following the procedure outlined in Appendix 3, we first search for possible steady solutions to (7). By definition, such solutions retain the same value as the years pass. Mathematically, we seek solutions of the form

$$x_i = x, \quad x \text{ constant (independent of } i).$$

Substituting into (7), we obtain

$$x = rx(1-x), \quad \text{so that either } x = 0 \text{ or } 1 = r(1-x), \text{ i.e., } x = 1 - r^{-1}. \tag{8}$$

Not surprisingly, according to our model the populations can persist forever only in a state of extinction $(x=0)$ or at the single nonzero level $(x=1-r^{-1})$ at which the population exactly replaces itself – because the birthrate $r(1-x)$ is unity. Note that the steady state $x=1-r^{-1}$ is biologically irrelevant when $r<1$, for then $x<0$.

The natural question now arises: To which of the two possible steady states, if either, does the population tend as the years go by? For lack of the appropriate mathematical tools, we must be content for the moment with supplying an answer to the following somewhat less ambitious question: If the population level starts *near* one of the steady states, what will be its subsequent behavior (at least for a while)?

Because x represents the steady state, the mathematical translation of the phrase "the population is near its steady-state level" is

$$x_i = x + y_i, \quad y_i \text{ small.} \tag{9}$$

As in Appendix 3, we exploit the smallness of y_i by approximating the right side of the governing equation (7), using Taylor series. As a result, we obtain the following equation [which is equivalent to (A3.33), in the present notation]:

$$y_{i+1} = f'(x) y_i, \tag{10a}$$

where $f'(x) \equiv df(x)/dx$. Here,

$$f(x) \equiv rx(1-x); \quad f'(x) = r(1-2x). \tag{10b}$$

We have gone as far as we can in a general fashion. To complete the stability analysis we must consider the different steady states separately.

Stability of $x=0$. For this case, (10) yields

$$f'(0) = r, \quad y_{i+1} = ry_i. \tag{11}$$

This steady state is thus stable for $0 < r < 1$ (the perturbation y_i approaches zero as $i \to \infty$, so that the solution itself approaches the given steady-state point) and unstable for $r > 1$.

Stability of $x = 1 - r^{-1}$. In the present case, (10) specializes to

$$f'(1 - r^{-1}) = 2 - r; \quad y_{i+1} = (2-r) y_i. \tag{12}$$

Because

$$-1 < 2 - r < 1 \quad \text{when } -3 < -r < -1,$$

we have stability for $1 < r < 3$ (monotonic for $1 < r < 2$, oscillatory for $2 < r < 3$) and instability otherwise (monotonic for $0 < r < 1$ and oscillatory for $3 < r$). (See Figure A3.1.)

Conjectured global behavior

Our findings concerning steady states and their stability are summarized in Figure 2.1. Based on these findings, we shall now make some conjectures about the general overall or *global* behavior of all possible solutions

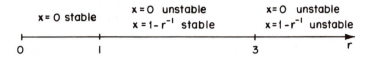

Figure 2.1. Steady states and their stability for various values of the maximum growth rate r.

(i.e., solutions for all possible initial conditions) to the governing equation (7), for various values of the parameter r. We shall be guided by the following principles:

I If the solution approaches any finite fixed population level as $i \to \infty$, the value approached must be one of the steady states [see equation (A3.27)].

II If the solution does not approach a finite steady state, then either it must vary continually while remaining bounded or it must grow indefinitely large.

III Conjectured overall behavior must be consistent with the behavior we have found for the solutions in the neighborhoods of steady states.

IV Biological intuition will be helpful in our conjectures, but we must constantly bear in mind that we are attempting to forecast the behavior inherent in a certain mathematical model, not (for the moment) in a genuine biological situation with all its complexity.

V Solutions of the governing equations for some particular cases can and should be used to test our conjectures. A hand calculator often provides sufficient computer power for such a test. (For a sample program, see the supplement to this chapter.)

When $0 < r < 1$, there is a single stable steady state, $x = 0$. Because r is the largest possible birthrate [see equation (4)], the organism cannot replace itself, and extinction must result. We thus conjecture that all solutions to (7), whatever the initial condition, tend to $x = 0$. [We must avoid the biologically unreasonable initial conditions for which the population level is negative – see Exercise 1(a).] This conjecture is supported by some sample computations (Figure 2.2).

When $1 < r < 3$, there is also a single stable steady state, and it is natural to conjecture that here, too, all solutions ultimately approach this state. This conjecture is supported by the fact that if the population begins below the steady state, then the growth rate is greater than unity, and the population increases, whereas populations that begin above the steady state will decrease in the next year (Figure 2.3). Again, sample computations bear out the conjecture (Figure 2.4).

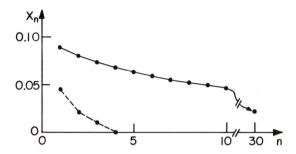

Figure 2.2. Hand-calculator solution of (7) for $x_0 = 0.9$ (not shown). Broken line: $r = 0.5$. Solid line: $r = 0.995$. Population tends to extinction as expected.

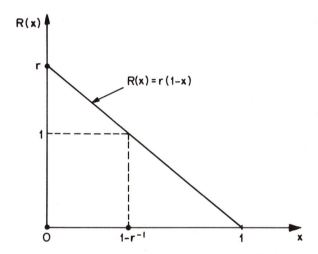

Figure 2.3. A graph of the growth rate R according to the model equation (7), when $r > 1$. Heavy dots represent possible steady-state values $x = 0$ (extinction persists) or values of x where $R = 1$ (the population just reproduces itself). Note that when x is greater (less) than the steady state, the growth rate is less (greater) than unity.

An important observation is that the decay of perturbations to the steady state is oscillatory when $2 < r < 3$. As an example, take $r = 2.5$, in which case the steady state is $x = 0.6$. If $x_0 = 0.7$, above the steady state, then $x_1 = 0.525$, *below* the steady state, albeit closer to it (Figure 2.5). For $2 < r < 3$, the decrease for populations that are above the steady state is so great that in the following year they are thrown below the steady state, only to surpass the steady state in the following year. When $r > 3$, the

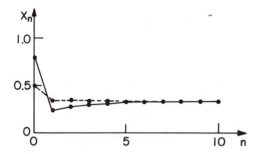

Figure 2.4. Solution of (7) for $r=1.5$, $x_0=0.5$ and 0.8. Populations that start near the steady state $x=\frac{1}{3}$ are predicted to approach the steady state monotonically, and this is borne out by the simulation. Note, however, that the large initial population drops below the steady state before its monotonic approach.

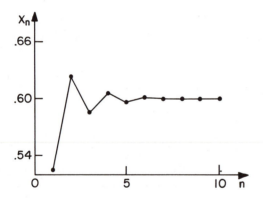

Figure 2.5. Solution of (7) for $r=2.5$, $x_0=0.7$ (not shown). As predicted, the steady state $x=\frac{3}{5}$ is approached in an oscillatory fashion.

effect is so strong that the steady state is actually unstable: If the population starts near the steady state, it will oscillate with ever increasing amplitude according to the linearized equation (12). When the oscillation amplitude ceases to be small, we have no detailed information on future behavior, as the linearization that resulted in (12) is no longer a valid approximation. But because there is no stable steady state, it must be that the population level continues to vary as time passes.

Our conjectures about the overall behavior of the solutions to (7) are summarized in Figure 2.6.

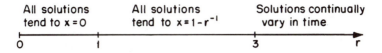

Figure 2.6. Conjectured qualitative behavior for various values of the maximum growth rate *r*.

The role of parameters

Figure 2.6 provides a characteristic illustration of the role of a parameter (in this case *r*) in mathematical models. A parameter is fixed for a given realization of the model [we perform computer simulations of (7) for fixed values of *r*, $i = 0, 1, 2, \ldots$]. But we wish to understand the behavior of the model for a whole range of parameter values. As will be seen in later chapters to be completely typical, Figure 2.6 shows that *the model yields different modes of qualitative behavior for different sets of parameter values.*

There are transition values of the parameter that separate one domain of behavior from another. (For example, in Figure 2.6 the transition value $r = 1$ separates cases in which the organism becomes extinct from cases in which it settles down to a steady population level.) Generally, there is little interest in the predicted model behavior when the parameter is precisely equal to a transition value. The reason is that the model is very sensitive to slight changes when a parameter is at or near such a value. In practice, parameters can generally be estimated only roughly, so that such sensitive conclusions are unreliable. Moreover, in such sensitive parameter regions the inevitable simplifications in a model are particularly influential – which is another reason why it is usually not worthwhile to calculate behaviors for transition values of parameters.

Control oscillations

Returning to our original biological problem, we have seen that the highly oversimplified case of constant growth rate becomes modified in a reasonable way when account is taken of a possible decrease in growth rate that arises from an increase in population density. A population that starts with a high growth rate will not grow indefinitely, according to our model. The increase in population brings about a decrease in growth rate, which leads in turn to a reversal of the growth trend. One can think of nature as equipped with a built-in control mechanism to keep popula-

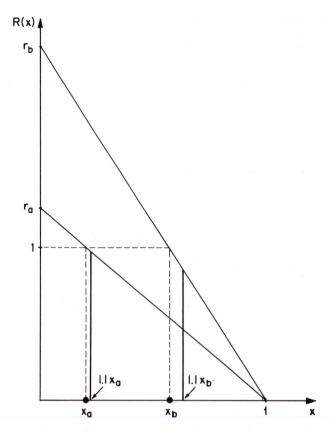

Figure 2.7. Growth rates (heavy vertical lines) corresponding to two different maximum growth parameters r_a and r_b, $r_b > r_a$, in both cases for population levels that are 10% above their steady values (x_a and x_b, respectively). Note that the larger drop in the growth rate corresponds to the larger maximum growth parameter.

tions within bound. Especially interesting in this regard is the situation in which r is large. Here the growth-rate curve has a very steep slope. The control is "tight" – there is a large response to minor deviations (Figure 2.7). Indeed, the control mechanism can overreact, as we have seen, leading to the opposite of the desired effect – oscillations of population that diverge from the steady-state value.

The phenomenon illustrated here is quite general – overcontrol can lead to oscillations that may even diverge. The phenomenon is exemplified in beginning drivers who overcorrect for deviations in course, leading to an oscillating path. In the problem under consideration there is a lag of a year between cause and effect (change in population, change in number of births). If the lag between information and action is longer, then the phenomenon of overcontrol is more likely (Exercise 5). This aspect of the phenomenon is also general. Imagine how difficult it would be to drive an automobile if there were a delay of one second while your mind processed what your eye perceived.[1]

Oscillations have been observed in biochemical systems, such as that for cAMP signaling in the cellular slime mold (Chapter 6) and that for glycolysis. [A major reference for the theoretical study of biological oscillations is Winfree (1980).] It has been conjectured that in some instances oscillations might have first arisen when control loops became too long (increasing destabilizing delay) or too tight [see Chapter 3, by P. Rapp, in Segel (1980)].

The period-doubling route to chaos[2]

We shall now summarize very briefly, for more advanced students, what has recently been discovered concerning the behavior of solutions to equation (7) when $r > 3$, so that there are no stable steady solutions. These results are taken from articles by May (1974, 1975) that can be consulted for details of the analysis.

(a) It can be shown that for $3 < r < 3.449$, a solution of period 2 exists and is stable to small perturbations. (By a solution or *cycle* of period n, we mean a set of values of x_i that repeats exactly every n years.) Indeed, for all permitted initial conditions, population levels tend closer and closer to an exact repetition every two years (Figure 2.8).

(b) For $3.449 < r < 3.544$, solutions tend to a stable cycle of period 4 [Exercise 2(b)].

(c) For $3.544 < r < 3.570$, there is a succession of transitions between stable solutions of period 8, period 16, and so forth. That is, there are narrower and narrower ranges of the parameter r corresponding to which populations tend to solutions of longer and longer even periods. One speaks of a **period-doubling phenomenon**.

(d) For $r > 3.570$, the behavior of solutions depends critically on initial conditions. Starting from uncountably many initial points, the solution never settles down to any period, but rather continues to change in a seemingly random way. Moreover, for $r > 3.828$ there are initial points

1 It has recently been shown in ecological contexts that delay can in some cases be stabilizing, when there are resonances with the intrinsic period of a phenomenon (Cushing and Saleem, 1982; Cooke and Grossman, 1982).
2 Material from here to the end of this chapter is more advanced and may be skipped.

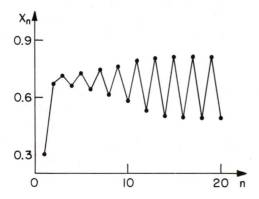

Figure 2.8. Solution of (7) for $r = 3.25$, $x_0 = 0.9$ (not shown). Population levels eventually oscillate with a two-year period.

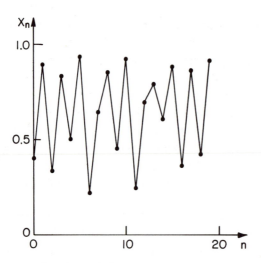

Figure 2.9. Solution of (7) for $r = 3.75$, $x_0 = 0.4$. No regular pattern can be discerned in this "chaotic" parameter regime.

corresponding to which the solution is cyclic of period n for any integer n, even or odd. In practice, round-off errors in the computer calculations (and disturbances in nature) perturb the developing solution so that its course probably will seem quite random in all cases. In this parameter range, then, solutions are said to be **chaotic** (Figure 2.9).

The preceding findings are summarized in Figure 2.10, which can be regarded as a more detailed look at the right portion of Figure 2.6. Note

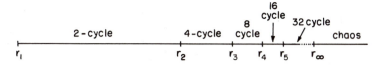

Figure 2.10. Schematic representation of the period-doubling route to chaos. Critical values of a parameter, r_n, divide regions in which solutions tend to cycles of period 2^{n-1} from regions in which solutions tend to cycles of period 2^n, $n = 1, 2, \ldots$. As $n \to \infty$, $r_n \to r_\infty$. For $r > r_\infty$, solutions are "chaotic." [For (7): $r_1 = 3$, $r_2 = 3.449$, $r_3 = 3.544, \ldots, r_\infty = 3.570$.]

that we have denoted by r_n the parameter value at which the nth period doubling takes place.

The qualitative results obtained are by no means restricted to the model represented by equation (7). They have been demonstrated for all models of the form $x_{i+1} = f(x_i)$ in which f is hump-shaped, such as $f(x) = rx(1-x)$ of equation (7). Indeed, the "period-doubling route to chaos" probably occurs in a wide variety of contexts. The following "universal relation" has been shown to hold in all "humped" cases between three successive values of the parameter at which period doubling occurs:

$$\lim_{n \to \infty} \frac{r_{n+1} - r_n}{r_{n+2} - r_{n+1}} = 4.6692016\ldots. \tag{13}$$

Usually the limit is closely approached even when n is not large. For example, for the case we have discussed, equation (13) is accurate to within about 1% even when $n = 1$:

$$\frac{r_2 - r_1}{r_3 - r_2} = \frac{3.449 - 3.000}{3.544 - 3.449} = 4.73.$$

For further information, see Hofstadter (1981).

The implications of these results might be viewed in two ways. An optimist might say that behavior that seems unpredictable can follow from laws of interaction that are simple and hence discoverable. A pessimist would perhaps stress the fact that even equation (7), almost the simplest mathematical model one might conceive, can under certain circumstances yield unpredictable dynamic behavior. Such pessimistic views were first put forward in a meteorological context to argue that long-term weather forecasts are an impossibility. It might similarly be argued that long-term forecasts of ecological implications may be impossible.

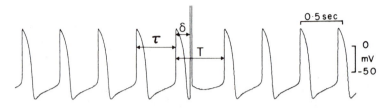

Figure 2.11. Time course of the transmembrane electrical potential from an aggregate of embryonic heart cells. Left: Spontaneous pulses. Right: After administration of a brief depolarizing stimulus (off-scale response) δ msec after the graph sharply rises, the spontaneous-state period τ is shifted to a new value T. [From Guevara et al. (1981), *Science,* 214: 1350–3, Figure 1A. Copyright 1981 by the AAAS.]

Chaotic dynamics in cardiac cells

The rich dynamic behavior that can be exhibited by difference equations has recently been shown to be relevant to some interesting experiments in cellular biology (Guevara, Glass, and Shrier, 1981). It has been shown that specially prepared embryonic chick heart cell aggregates approximately 100 μ in diameter will spontaneously beat with a period of order 1 sec. Microelectrodes have revealed the presence of corresponding periodic electrical impulses. Their rise-fall-recovery shape is strongly reminiscent of the "action-potential" voltage pulses that travel along axons (the long projections from nerve cells) (Figure 2.11).

If the preparation is periodically stimulated via a current pulse through a microelectrode, the nature of the response depends on the interstimulus interval. For example, under certain conditions the stimulus evokes a single responding impulse after a certain fixed delay time. A slight change in conditions will only alter the delay. But under other conditions, following successive stimuli there is a slow lengthening of the delay, until finally a response is skipped. General behavior of this type, which may be either regular or irregular, is very like the clinically observed "Wenckebach phenomenon" (Figure 2.12).

Analysis of these results has been carried out by means of a basic phase-resetting experiment in which one measures the shift in a signal induced by a solitary stimulus as a function of the time of stimulus application. To explain these experiments more precisely, let us designate the sharp rise in the graph as the onset of the periodic signal. Let δ be the time interval after onset at which the stimulus is delivered, and let T be the corresponding total duration of the signal, from the original onset to

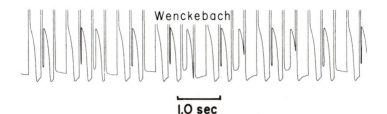

1.0 sec

Figure 2.12. Transmembrane recording illustrating an irregular beat skipping that is characteristic of the clinically observed Wenckebach phenomenon. [From Guevara et al. (1981), *Science,* 214: 1350–3, Figure 1C(i). Copyright 1981 by the AAAS.]

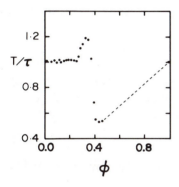

Figure 2.13. The function *g* defined in (15), as experimentally determined for embryonic chick heart cell aggregates. [From Guevara et al. (1981), *Science,* 214: 1350–3, Figure 1D(i). Copyright 1981 by the AAAS.]

the next onset (Figure 2.11). The inter-onset time interval T will in general depend on the stimulus time δ. To express this, it is convenient to use the **phase** ϕ rather than δ itself, where ϕ is the ratio between δ and the undisturbed period τ of the signal:

$$\phi \equiv \delta/\tau, \quad 0 \leqslant \phi < 1. \tag{14}$$

One also employs τ to normalize T, so that the results of the phase-setting experiments consist of a graph of the function g defined by

$$T/\tau = g(\phi). \tag{15}$$

Figure 2.13 shows one of the experimentally determined functions g.

A major experimental effort has been devoted to examining the response of the cell aggregate to a train of impulses separated by a uniform

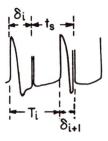

Figure 2.14. Graphical demonstration of (16), in the case in which $T_i < \delta_i + t_s < T_i + \tau$.

time interval t_s. Consultation of Figure 2.14 leads to the conclusion that the times after onset of successive pulses, δ_{i+1} and δ_i, are related to the duration T_i by

$$\delta_{i+1} + T_i = \delta_i + t_s. \tag{16}$$

It is assumed that the measured function g of (15) continues to give approximately correct results even for one of a train of stimuli. Division of (16) by τ then yields the phase relationship

$$\phi_{i+1} = \phi_i - g(\phi_i) + (t_s/\tau). \tag{17}$$

(There are circumstances in which this relationship must be modified. See Exercise 9.)

Equation (17) is a recursion relation of the prototypical form (1) with which we began this chapter:

$$\phi_{i+1} = f(\phi_i), \quad f(\phi) = \phi - g(\phi) + t_s/\tau. \tag{18}$$

Figure 2.15 depicts the graph of $f(\phi)$, constructed from measurements of g. Guevara, Glass, and Shrier (1981), relying on mathematical analysis by Guevara and Glass (1982), discuss the good correspondence between observation and predictions obtained by analyzing equation (17). [Strictly speaking, the generalization of (17) in Exercise 9 was analyzed.] In particular, there is qualitative correspondence between theory and experiment with respect to regions of complicated, irregular dynamics.

Cell maturation

As one simple model for stem-cell–mature-cell transition in the generation of red and white blood cells and platelets, Mackey and Glass (1977) proposed the delay differential equation

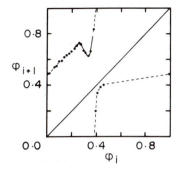

Figure 2.15. Experimentally determined graph of the function f defined in (18), in terms of the function g of Figure 2.13. [From Guevara et al. (1981), *Science,* 214: 1350-3, Figure 2B. Copyright 1981 by the AAAS.]

$$\frac{dP(t)}{dt} = \frac{\beta\theta^n [P(t-\tau)]^n}{\theta^n + [P(t-\tau)]^n} - \gamma P(t). \tag{19}$$

Here, $P(t)$ is the number of mature cells circulating in the bloodstream. The first term on the right side of (19) represents cell production in the bone marrow, and the second term represents cell death. The model incorporates a fixed delay τ between the commencement of cell production and the actual appearance of the cells in the blood. (The quantities β, θ, n, γ, and τ are constants.) Mackey and Glass (1977) reported that analysis of (19) reveals period doubling and then chaos for increasing values of τ, in close analogy to the behavior of the difference equations such as (7). And, indeed, various kinds of regular and irregular periodicities in blood count have been reported in the clinical literature.

The general subject of "chaos" has only recently become an object of intensive investigation by theoreticians. The cited studies on chaotic effects involving cardiac and blood cells allow one to conjecture that such effects may be of considerable importance in certain areas of molecular and cellular biology.

Exercises

1. (a) Given that $x_i > 0$, $r > 0$, show that in (7), $x_{i+1} < 0$ only if $x_i > 1$.
   (b) Show that (7) permits a value of x_{i+1} that is greater than unity only if $r > 4$, assuming $0 < x_i < 1$.
   (c) What conditions on x_0 are necessary and sufficient to guarantee $x_i > 0$, $i = 1, 2, \ldots$.
2. (a) Repeat the calculations of Figures 2.2, 2.4, 2.5, 2.8, and 2.9, but in each case employ a different value of r that is predicted to give the same qualitative behavior. Also repeat the calculations using the same r but a different initial condition.

†   (b) When $r = 3.5$, show for at least two different initial conditions that the solution approaches a situation that repeats itself every four years.

†   3. Consider the equation

$$x_{n+1} = \tfrac{1}{4}x_n(1 + 4x_n - x_n^2).$$

(a) Find the steady states.

(b) Examine the stability of each steady state.

(c) Conjecture the qualitative behavior of the solution. Test your conjecture with the aid of a hand calculator.

    4. Consider the equation

$$x_{i+1} = x_i \exp[s(1 - x_i)], \quad s > 0. \tag{20}$$

(a) Show, using Taylor series, that the birthrate decreases linearly with x_i for x_i small, as in (7). But (20) has the advantage that the postulated birthrate never becomes negative.

†   (b) Find the steady states of (20) and examine their stability.

(c) Equation (20) fits Figure 2.10 (with s replacing r) when $s_0 = 2$, $s_1 = 2.526$, $s_2 = 2.656$, $s_3 = 2.692$ (May, 1975). Use a hand calculator to check in various special cases that solutions behave as expected.

(d) Taking into account the effects of population level on natural selection, May (1979) considered an example in which

$$x_{n+1} = ax_n^3 + (1 - a)x_n, \quad 0 < a < 4.$$

Show that the three steady states of this equation are $x = 0$ and $x = \pm 1$. Examine the stability of the three steady states when $a = \tfrac{1}{2}$. Conjecture the qualitative behavior of x_n. Test your conjectures by computer calculations.

The next two exercises deal with the effect of delay in population regulation. Mathematically, the problems shift from first- to second-order difference equations. An understanding of Example 2 in Appendix 3 is required, which in turn is based on some aspects of complex numbers that are explained in Appendix 6.

    5. When animals feed on plants, the food available for the animals in a given year, and hence their birthrate, may depend on how much was eaten during the previous year. This motivates modifying (7) to

$$x_{i+1} = rx_i(1 - x_{i-1}). \tag{21}$$

(a) Show that the steady states of x of (21) are exactly the same as those of its undelayed counterpart (7).

(b) Introduce the perturbation $y_i = x_i - x$. Find the equation satisfied by y_i. Delete nonlinear terms, and find the resulting linear equation for small perturbations y_i. Show that for the positive steady state, this equation can be written

$$y_{i+1} = y_i - (r - 1)y_{i-1}. \tag{22}$$

(c) Assuming a solution to (22) of the form $y_i = k^i$, show that

$$k = \tfrac{1}{2} \pm \tfrac{1}{2}\sqrt{(5 - 4r)}.$$

(d) Show that when $r < 1$, the positive steady state is unstable, whereas $x = 0$ is stable.

(e) Show that when $1 < r < \tfrac{5}{4}$, small perturbations to the positive steady state decay monotonically.

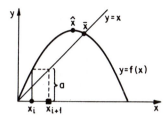

Figure 2.16. Diagram of the graphical iteration method discussed in Exercise 7.

(f) Find the general solution to (22) when $r > \frac{5}{4}$.

(g) Deduce from (f) that small perturbations to the positive steady state decay in an oscillatory fashion when $\frac{5}{4} < r < 2$ and grow when $r > 2$. (Comparison with Figure 2.1 shows that these results bear out the assertion that delay is generally destabilizing.)

(h) Explore numerically various values of $r > 2$. Can you find values of r for which the solution has period 2? Period 4? Apparent chaos?

6. Carry through an analysis similar to that of the previous exercises for the delayed version of (20):

$$x_{i+1} = x_i \exp[s(1 - x_{i-1})]; \quad s > 0.$$

7. (a) Given the relation $x_{i+1} = f(x_i)$ of (7), and given the location of x_i as shown in Figure 2.16, show that the point labeled with the coordinate x_{i+1} is correctly labeled.

(b) From part (a), show that points $x_0, x_1, x_2, \ldots$ can be successively found by the following geometric algorithm:
1. Start at x_0.
2. Move vertically to the graph of $f(x)$.
3. Move horizontally to the graph of $y = x$.
4. Go back to step 2.

(c) Deduce that steady solutions of (7) are located at the intersection of $y = x$ and $y = f(x)$. What is the geometric interpretation of the condition on $f'(x)$ that is necessary and sufficient for the stability of a steady solution?

(d) Make several careful large graphs of $f(x) = 2.5x(1-x)$. (Prepare one graph and trace the rest.) Starting with different values of x_0 and using the graphic algorithm of (b), verify that the successive x_i's approach the steady solution.

8. (a) Repeat part (d) of Exercise 7 when $f(x) = 3.25x(1-x)$. Now you will see that the steady solution is not approached.

(b) According to Figure 2.8, in this case the solution settles down to a cycle of period 2. This means that the *two-year function* $f_2(x) = f[f(x)]$ will have a steady-state point. Show that for $f(x) = rx(1-x)$, $f_2(x) = r^2(x - x^2)(1 - rx + rx^2)$.

(c) Rather than deal with the particular $f_2(x)$ just determined, it is more elegant to show some general properties of f_2. We shall suppose that $f(x)$ has a single maximum, say at $x = \hat{x}$, and that it is symmetric about $\hat{x}$. Show that these properties hold for $f(x) = rx(1-x)$. Given the concave downward shape of f, find a condition on $f'(0)$ that will guarantee that $f(x) = x$ has one positive solution, $\bar{x}$. We shall assume that this condition holds. (The points $\hat{x}$ and $\bar{x}$ are labeled in Figure 2.16.)

† (d) Show that $f_2(x)$ has a zero derivative at $\hat{x}$, and also at points $\bar{x}$ such that $\hat{x}=f(\bar{x})$. Deduce graphically that $f_2(x)$ has a zero derivative at precisely three points by showing that there are two values of $\bar{x}$, which we shall call $\bar{x}_1$ and $\bar{x}_2$.

† (e) Show that with the foregoing assumptions, f_2'' is positive at $\hat{x}$ and negative at $\bar{x}_1$ and $\bar{x}_2$. Deduce a two-humped shape for $f_2(x)$. Show that when $\bar{x}$ is unstable, then $y=x$ intersects $f_2(x)$ once at $\bar{x}$ and also at two other points, which we shall call $\bar{x}_1$ and $\bar{x}_2$.

(f) Because $\bar{x}_1$ and $\bar{x}_2$ are not steady points of f itself, it must be that $\bar{x}_1 = f(\bar{x}_2)$, $\bar{x}_2=f(\bar{x}_1)$. This is the solution of period 2. Although it would take a while, the reader might enjoy making a large graph of $f_2(x)$ when $f(x)=3.25x(1-x)$ and observing graphically the convergence of the solution to this 2-cycle. Graphs of this and more complex kinds can be found in the article by Hofstadter (1981) and in an excellent expository article by one of the leading researchers in this field, M. Feigenbaum (1980).

9. (a) Show, with the aid of a sketch, that (16) must be replaced by

$$\delta_{i+1}+T_i+\tau=\delta_i+t_s \quad \text{when } T_i+\tau<\delta_i+t_s<T_i+2\tau.$$

(b) By examining additional situations, give a convincing argument that, in general,

$$\phi_{i+1}=[\phi_i-g(\phi_i)+(t_s/\tau)+1] \quad (\text{mod } 1)$$

where $a=b$ (mod c) means that $a=cn+b$, $0 \leqslant b<c$, for some integer n. That is, b is the remainder when a is divided by c. [For example, we normally give the time in hours (mod 12), although sometimes we use hours (mod 24).]

Supplement: A program to iterate equation (2.7)

We present here a program for a hand calculator to provide successive iterates defined by equation (2.7):

$$x_{i+1}=rx_i(1-x_i), \quad i=0,1,2,\ldots. \tag{1}$$

For definiteness, the program is written for a particular calculator, the Texas Instruments TI Programmable 57, but the general principles (which are outlined in the following comments) are readily implementable on any particular calculator. The main object of this supplement is to demonstrate how quickly one can set oneself up to perform numerical experiments with equations like (1).

The general plan will be to place the desired value of the parameter r in storage register 2 and x_0 (the initial value) in storage register 1. As the iterations proceed, successive "answers" $x_1, x_2, \ldots$ will also be placed in storage register 1 – to serve as starting values for the forthcoming iteration. In the given program steps, each boxed symbol designates the required pressing of the corresponding button on the calculator.

Program step	Comments
LRN	Learn (LRN): Start teaching the calculator (i.e., begin the program).
STO 1	Place the number on display (which will be the result of the last iteration) in storage (STO) register 1.
(1 − RCL 1)	Recall (RCL) the number (x_i) from STO 1 and subtract it from unity [i.e., form $(1-x_i)$].
× RCL 1 × RCL 2	Multiply by x_i and by r [i.e., form $rx_i(1-x_i)$].
=	Put the result, which is x_{i+1}, in display.
R/S	Stop the program.
RST	Reset (RST): Set the calculator back to the beginning of the program.
LRN	End of program.

The program is complete. We now show how to run it, to obtain x_1, x_2, x_3 when $r=2.1$, $x_0=0.4$.

Program step	Comments
2 . 1 STO 2	Store $r=2.1$ in storage register 2.
RST	Position the computer at the beginning of the program.
0 . 4	Put the initial condition $x_0=0.4$ into the display.
R/S	Run the program; $x_1=0.50400$ appears in the display (we keep five significant digits).
R/S	Run the program again; $x_2=0.52497$ appears in the display.
R/S	Compute $x_3=0.52369$.

The iterate x_5 and all succeeding iterates equal 0.52381, which (to this accuracy) is exactly equal to the predicted steady-state value $1-(2.1)^{-1}$.

3

Population genetics: separate generations

This chapter is intended to give a flavor of how mathematical models can be helpful in understanding some of the many interacting factors that influence natural selection.[1] The mathematical ideas are very similar to those used in the first part of Chapter 2. In principle, either Chapter 2 or Chapter 3 can be skipped, but repetition of the same theme in a different biological context will reinforce command of the mathematical methodology. Also, Chapter 3 is a little more difficult than the main part of Chapter 2 (excluding the discussion of chaos) in that more manipulation of inequalities is required, and domains of different qualitative behavior are found in a two-dimensional parameter plane rather than a line.

Formulation

Let us consider **diploid** organisms (like humans) in which each gene is present in two copies. Each copy has two possible states or **alleles**. We shall concentrate on a trait that is determined by a single location on the chromosome. If we denote the two alleles by A and a, this means that as far as the gene is concerned there are three possibilities (**genotypes**), namely the **homozygotes** AA and aa and the **heterozygote** aA.

Two organisms with the same genotype may not have exactly the same characteristics or **phenotypes** (as in "identical" twins), and natural selection is traditionally held to operate on the phenotype. As is common in introductory treatments, however, we shall ignore the distinction between genotype and phenotype. We characterize the **viabilities** Φ_{AA}, Φ_{Aa}, and Φ_{aa} of the three genotypes as the average fraction of fertilized eggs that survive to reproduce. We assume that each genotype produces the same number of offspring (but see Exercise 7).

Consider the nth generation. Let p_n be the proportion of alleles A in the adult population, and let $q_n = 1 - p_n$ be the proportion of alleles a. Each reproductive cell (**zygote**) receives one allele from its mother and

---

1 Our discussion is inspired by Maynard Smith's treatment of population genetics in his valuable presentation of *Mathematical Ideas in Biology* (1968).

Table 3.1. *Changes in genotype balance assuming* N *births in* n*th generation*

Genotype	Number of zygotes in nth generation	Number of surviving adults in generation $n+1$
AA	Np_n^2	$\Phi_{AA} Np_n^2$
Aa	$2Np_n q_n$	$2\Phi_{Aa} Np_n q_n$
aa	Nq_n^2	$\Phi_{aa} Nq_n^2$

one from its father. Suppose that alleles occur at random (assuming that there are no mating preferences that depend on genotype).

Our assumptions imply that p_n is the probability that the mother donates an A, and q_n is the probability that the father donates an a. Thus, a fraction of choices $p_n q_n$ will provide an A from the mother and an a from the father. A fraction $q_n p_n = p_n q_n$ will provide an a from the mother and an A from the father. Consequently, a fraction $2p_n q_n$ of the zygotes will be of type Aa; that is, $2p_n q_n$ is the probability that of a pair of alleles one is A and the other is a. The probabilities of aa and AA are q_n^2 and p_n^2. Multiplication by the respective survival fractions yields the information on the zygote genotype distribution that is recorded in Table 3.1.

Using the last column of Table 3.1, we see that the total number of A alleles in generation $n+1$ (counting two A's for an AA and one for an Aa) is

$$2\Phi_{AA} Np_n^2 + 2\Phi_{Aa} Np_n q_n. \tag{1}$$

The total number of all alleles in the $(n+1)$st generation is

$$2N(\Phi_{AA} p_n^2 + 2\Phi_{Aa} p_n q_n + \Phi_{aa} q_n^2). \tag{2}$$

Thus, if p_{n+1} denotes the proportion of A alleles in generation $n+1$, then

$$p_{n+1} = \frac{\Phi_{AA} p_n^2 + \Phi_{Aa} p_n q_n}{\Phi_{AA} p_n^2 + 2\Phi_{Aa} p_n q_n + \Phi_{aa} q_n^2}. \tag{3}$$

We can simplify the preceding expression by introducing abbreviations for the ratios of fitnesses. It is conventional to write

$$\frac{\Phi_{AA}}{\Phi_{Aa}} = 1 - K, \quad \frac{\Phi_{aa}}{\Phi_{Aa}} = 1 - L. \tag{4a,b}$$

Because the ratios in (4) are nonnegative, the "relative selection coefficients" K and L satisfy $K \leqslant 1$, $L \leqslant 1$. In the text we shall restrict ourselves to the cases

$$K < 1, \quad L < 1. \tag{5a,b}$$

We also leave aside the situations in which two genotypes are equally fit; that is, we assume that

$$K \neq 0, \quad L \neq 0. \tag{5c,d}$$

On substitution of (4) into (3), a little algebra (Exercise 1) yields

◆◆
$$p_{n+1} = \frac{p_n(1 - p_n K)}{1 - K p_n^2 - L(1 - p_n)^2}. \tag{6}$$

Steady-state solutions

It would be helpful to be able to find an explicit expression for p_n in terms of K and L, but no general method for solving this type of equation is known. Nonetheless, we shall be able to find quite a bit of information about the solution.

The first step is to see if there are any **steady-state** or **equilibrium solutions** to (6). Such solutions keep the same value for all generations; that is, they satisfy

$$p_n = p \quad \text{for all } n, \text{ with } p \text{ a constant.} \tag{7}$$

Substituting (7) into (6), we find that

$$p = \frac{p(1 - pK)}{1 - Kp^2 - L(1 - p)^2}, \tag{8}$$

or (Exercise 1)

◆◆
$$p = 0, \quad p = 1, \quad p = \frac{L}{K + L}, \tag{9a,b,c}$$

assuming that $K + L \neq 0$. If $K + L = 0$, there is no separate solution (9c).

By their definition, gene frequencies must lie between zero and one. For the solution (9c) to be biologically meaningful, and also to be distinct from $p = 0$ or $p = 1$, we must have

$$0 < L/(K + L) \quad \text{and} \quad L/(K + L) < 1. \tag{10a,b}$$

To proceed, we shall multiply both sides of the inequalities in (10) by $K + L$. Giving consideration to the two possible signs of $K + L$, we find the following results:

$K+L>0$:   (10a) implies $L>0$; (10b) implies $L<K+L$ (i.e., $K>0$).

$K+L<0$:   (10a) implies $L<0$; (10b) implies $L>K+L$ (i.e., $K<0$).

Thus, if (9c) is to be a biologically meaningful distinct point, then either

$$K>0, \quad L>0 \quad \text{or} \quad K<0, \quad L<0. \tag{11a,b}$$

Stability theory

We now examine the stability of the steady states (9). This will enable us to determine whether a solution to (6) that starts in the zeroth generation *near* one of the steady states will tend as generations pass to come closer to this steady state or to diverge further from it.

From Appendix 3, if we define a perturbation δ_n from a steady state p by

$$p_n = p + \delta_n, \tag{12}$$

then, if δ_n is small, to good approximation it satisfies the equation

$$\delta_{n+1} = R\delta_n. \tag{13}$$

From equations (A3.33) and (6), the constant R is given by

$$R = \frac{\partial}{\partial p_n} \left(\frac{p_n(1-p_n K)}{1-Kp_n^2 - L(1-p_n)^2} \right)_{p_n=p}. \tag{14}$$

A calculation (Exercise 1) shows that

$$R = \frac{p^2(K+L-2KL) + 2Kp(L-1) + 1 - L}{[1-Kp^2 - L(1-p)^2]^2}. \tag{15}$$

Let us first examine the stability of the steady state $p=0$. Here, from (15),

$$R = \frac{1}{1-L}. \tag{16}$$

As is indicated in Figure A3.1, the steady state is stable if and only if $-1 < R < 1$, that is, if and only if

$$-1 < \frac{1}{1-L} \quad \text{and} \quad \frac{1}{1-L} < 1. \tag{17a,b}$$

Because $1-L>0$, (17a) is satisfied for all L (because all positive numbers are greater than -1). Inequality (17b) requires that $L<0$. By (4b), this in turn means that $\Phi_{aa} > \Phi_{Aa}$.

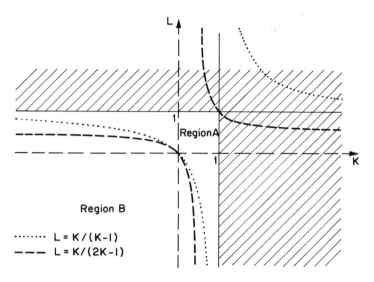

Figure 3.1. Diagram for determining the behavior of the function $R(K, L)$ given in (19). (There is no biological significance to values of K and L that correspond to points in the shaded region.)

Our stability result makes perfect sense. If and only if the *aa* homozygote is more fit than the heterozygote *Aa* will a population in which almost all alleles are *a* tend to a population containing *only* such alleles. (If *A* is rare, the fitness of the *AA* homozygote is irrelevant.) The fact that $R \equiv 1/(1-L) > 0$ means that approach or departure from equilibrium (of small perturbations) is always monotonic (see Figure A3.1).

Let us now examine the stability of the steady-state solution $p = 1$. From (15), we now obtain

$$R = 1/(1-K) \qquad (18)$$

instead of $R = 1/(1-L)$ as in (16). Thus, to describe the stability of the *AA* homozygote (stable when $K < 0$, unstable when $K > 0$), we merely substitute the fitness parameter K for the parameter L that governs the *aa* homozygote.

At the third possible steady-state point, $p = L/(K + L)$. Substitution into (15) gives, in this case,

$$R = (K + L - 2KL)/(K + L - KL). \qquad (19)$$

We now must determine when $R > 1$, $0 < R < 1$, and so forth (see Figure A3.1). In considering R, we must keep in mind that K and L cannot be greater than unity, by (5). Interest is thus confined to the unshaded portions of the K–L plane shown in Figure 3.1. Moreover, for the third steady-state point to be positive, we have seen in (11) that K and L must have the same sign. Hence, in examining R, we need only consider parameter values that fall in the regions marked A and B in Figure 3.1.

For what values of K and L is $R > 1$, and for what values is $R < 1$? In other words, when is the function $R(K, L) - 1$ positive, and when is it negative? From (19),

$$R(K, L) - 1 = -KL/(K + L - KL).$$

From this we see that $R(K, L) - 1$ changes sign when either K or L or $K + L - KL$ changes sign. Continuous factors such as K, L and $K + L - KL$ can change sign only by passing through zero. We are thus led to the conclusion, not only for the particular function $R(K, L) - 1$ but in general, that *a function can change sign only when it passes through zero* (numerator vanishes) *or through infinity* (denominator vanishes), *or when it makes a discontinuous finite jump.* (The last possibility is rarely met in practice.)[2]

In our particular case, $R - 1 = 0$ when $K = 0$ or $L = 0$, so that we must check for a sign change when the axes are crossed. In addition, $R - 1 = \infty$ when

$$K + L - KL = 0, \quad \text{i.e., when } L = K/(K - 1). \tag{20}$$

This curve is plotted in Figure 3.1 as a dotted line.

Consider any point in region B in Figure 3.1 (the third quadrant), for example, $K = -\frac{1}{2}$, $L = -\frac{1}{2}$. Here, $R = \frac{6}{5}$. Because $R > 1$ at this single point, $R > 1$ in all of region B (because $R - 1$ can change sign only when the K axis, the L axis, or a dotted curve is crossed, as we have seen). Thus, in all of region B, the steady-state point in question is unstable.

Consider the point $K = \frac{1}{2}$, $L = \frac{1}{2}$ in region A. Here $R = \frac{2}{3}$, so $R < 1$ throughout this region. To pursue the stability investigation, we must now determine whether or not $R > 0$ in region A. Proceeding as before, we search out the curves $R = 0$, $R = \infty$, where R may change sign. From (19) we see that the latter curve is given by (20) – and indeed it is no surprise that R and $R - 1$ both tend to infinity under the same conditions. We observe from (19) that $R = 0$ when

$$K + L - 2KL = 0, \quad \text{i.e., } L = K/(2K - 1).$$

As is shown in Figure 3.1, neither curve where R can change sign passes through region A (Exercise 2), so that R is either always negative

2 Questions of sign change will again be important in the study of the phase plane (see Appendix 5).

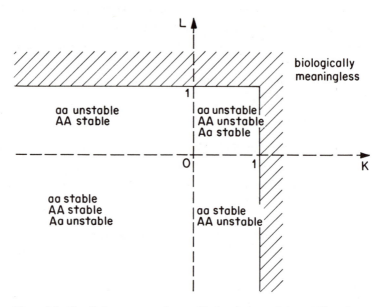

Figure 3.2. The *K-L* parameter plane, with descriptions of the stability of possible steady-state points.

or always positive in this region. In fact, the latter alternative is correct, for we already know that $R(\frac{1}{2}, \frac{1}{2}) = \frac{2}{3}$.

(Generally, to determine detailed stability behavior we must determine parameter regions wherein $R > 1$, $0 < R < 1$, $-1 < R < 0$, $R < -1$. In the present case we do not have to proceed further, for all the permissible parameter values turn out to be covered by the first two cases.)

It is convenient to display our findings about the stability of steady states by noting our results in different regions of the *K-L* plane. We shall refer to the three steady-state points of equation (9) as *aa*, *AA*, and *aA*, respectively. Figure 3.2 summarizes our conclusions.

Conjectured global behavior

Our calculations to this point have given us information about what is called **local behavior**, behavior in the "locality" of, or near, one of the steady-state points. Our ultimate goal is the determination of **global behavior**: We wish to discover the fate of the probability p_n as n increases, under all circumstances. Given our results so far, we can make an intelligent guess about the global behavior. Consider, for example, the upper left unshaded region in Figure 3.2, where $K < 0$, $0 < L < 1$. In

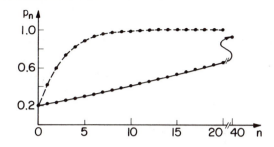

Figure 3.3. Hand-calculator solutions of equation (6) with $p_0 = 0.2$ for $K = -0.1$, $L = 0.1$ (solid line) and $K = -0.5$, $L = 0.75$ (dashed line).

terms of viabilities, the limits on K and L translate to $\Phi_{aa} < \Phi_{Aa} < \Phi_{AA}$. If there is a low proportion of A alleles in the population (p_n small), then most of the organisms will be of type aa or Aa, and the latter will be favored. Thus, the percentage of heterozygotes Aa will increase, but there will also be a growth in the number of AA's. These are the fittest of all, and they are expected to predominate as time goes on.

In the particular case we have just been considering, there is only one stable steady-state point, $p_n = 1$. Recalling from Appendix 3 that limiting points (if any) must be steady-state points, we anticipate that the system approaches this point as $n \to \infty$. To lend weight to our arguments, we should attempt some calculations for particular cases. Results of such calculations, carried out on a small hand calculator, are shown in Figure 3.3. They provide evidence that, indeed, p_n, the proportion of A, approaches unity as $n \to \infty$ when $K < 0$, $0 < L < 1$.

From a mathematical point of view, the case $K < 0$, $L < 0$ is particularly interesting, for two stable steady-state points are predicted. Here, $\Phi_{Aa} < \Phi_{aa}$, $\Phi_{Aa} < \Phi_{AA}$, so that the heterozygote is least fit. We expect that the genotype will approach either AA or aa as $n \to \infty$, depending on whether the initial proportion of A in the population is greater than or less than the value $L/(K+L)$ at the unstable steady state. Calculations for special cases confirm this conjecture for the case $K = -0.25$, $L = -0.75$, $L/(K+L) = 0.75$ (Figure 3.4).

Similar reasoning for the remaining cases leads from the calculated behavior of the steady-state points found in Figure 3.2 to the qualitative behavior described in Figure 3.5. We have not proved that this behavior actually occurs, but if our plausible general reasoning is backed up by sufficient computer simulations (see Exercise 3), then we may legitimately have considerable confidence in our inferences. See Figure 3.6 for another representation of our conclusions.

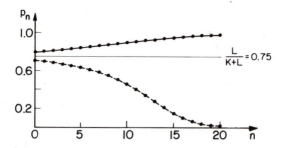

Figure 3.4. Hand-calculator solutions of equation (6) for $K = -0.25$, $L = -0.75$ with $p_0 = 0.8$ (solid line) and $p_0 = 0.7$ (dashed line). The horizontal line represents an unstable steady state.

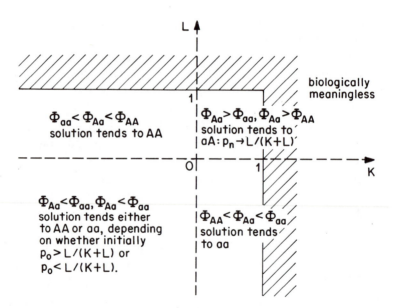

Figure 3.5. The K–L parameter plane, with descriptions of qualitative behavior.

Quantitative considerations

Until this point we have emphasized the less familiar and perhaps more important qualitative conclusions that can be drawn from our mathematical models. Now we shall illustrate with an example the type of quantitative inferences that can be drawn.

Example: Consider equation (6), with $K = 0.5$ and $L = 0.25$. How many generations will it take for p_n to change from an initial value of $p_0 = 0.01$ to within 1% of its asymptotic value? What if $p_0 = 10^{-6}$?

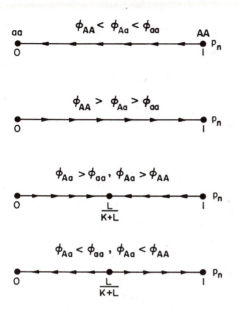

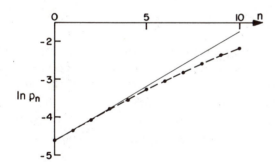

Figure 3.6. Qualitative behavior of the proportion p_n of A alleles in the nth generation, for various relations between the relative fitnesses. The four cases correspond to the four regions in Figure 3.5. (Black dots indicate steady states.)

Figure 3.7. Logarithmic plot of calculated solution p_n to equation (6) for $K = 0.5$, $L = 0.25$, $p_0 = 0.01$. The straight line is the prediction of linear theory for small p_n.

Solution: Computer iterations show that about 20 generations are required, when $p_0 = 0.01$, to closely approach the asymptotic value $L/(K+L) = \frac{1}{3}$ that is predicted in Figure 3.5; see Figures 3.7 and 3.8.

When $p_0 = 10^{-6}$, we can easily use analytical methods to calculate the additional time necessary for p_n to rise from 10^{-6} to 10^{-2}. During this whole period, $0 < p_n \ll 1$. Thus, we can employ the results from our investigation of the stability of the steady solution $p = 0$. Consequently, from (16) we use the approximate relation

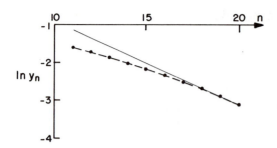

Figure 3.8. Logarithmic plot of calculated function $y_n \equiv p_n - \frac{1}{3}$, where p_n is the solution described in the legend for Figure 3.7. The straight line is the prediction of linear theory for a small departure y_n from the asymptotic state $p_n \to \frac{1}{3}$.

$$p_{n+1} = \frac{1}{1-L} p_n, \quad \text{so that } p_n = \left(\frac{1}{1-L}\right)^n p_0, \quad \ln p_n = n \ln\left(\frac{1}{1-L}\right) + \ln p_0.$$
$$\text{(21a,b,c)}$$

In the present case,

$$n = \ln \frac{p_n}{p_0} \bigg/ \ln\left(\frac{1}{1-L}\right), \quad \text{or } n = \frac{\ln 10^4}{\ln(4/3)} = \frac{9.21}{0.288} = 32. \qquad \text{(22a,b)}$$

Thus, only about 30 additional generations are required to raise p_n by the additional factor of 10^4. [Note, from Figure 3.7, that even when $p_n \approx 0.01$, the approximation

$$\ln p_n = 0.288n + \ln p_0 \qquad (23)$$

of (21c) is quite accurate.]

The example just given is typical of strong selection. The strongest selection occurs for a lethal allele, which is eliminated in a single generation. Less drastic but still strong selection can markedly modify genotypes in a few tens of generations, as our example illustrated. To see the effect of a typical weak selection, we may take $L = 10^{-3}$ in (21). The frequency of A will shift from $p_0 = 0.01$ to $p_n = 0.1$ in n generations, where, from (22a),

$$n = \frac{1}{L} \ln \frac{p_n}{p_0} = 10^3 \ln 10 = 2.3 \times 10^3 \text{ generations.} \qquad (24)$$

In deriving (24), we have employed the Taylor approximations (for small L)

$$\frac{1}{1-L} \approx 1+L, \quad \ln(1+L) \approx L. \qquad \text{(25a,b)}$$

See equations (A2.9) and (A2.13).

Darwin postulated that selection is the major force affecting gene frequencies. Another alternative is the continual action of mutation. To check the relative importance of this factor in the simplest possible way, let us denote by μ the fraction per generation of a alleles that mutate to A. To simplify our first model, we neglect back-mutations from A to a.

We shall write an equation in terms of the frequency of a alleles in the nth generation, q_n. A fraction $1 - \mu$ of the a's will remain from each past generation. Thus,

$$q_{n+1} = (1-\mu)q_n, \quad q_n = (1-\mu)^n q_0. \tag{26}$$

In (24) we estimated how long it would take selection to increase the fraction of A from 0.01 to 0.1. Now we consider the complementary problem of estimating how quickly mutation will shift the frequency of a from 0.99 to 0.9. We obtain, from (26),

$$n \ln(1-\mu) = \ln q_n/q_0. \tag{27}$$

Because μ will be small [so that we can employ the equivalent of (25b)],

$$n = \frac{1}{\mu} \ln \frac{q_n}{q_0} = \frac{1}{\mu} \ln \frac{0.99}{0.9} = 0.095\mu^{-1}. \tag{28}$$

Measurements indicate that μ is of magnitude 10^{-6} per generation. Comparison of (24) and (28) thus shows that mutation will accomplish in about 10^5 generations what weak selection will bring about in approximately one-hundredth of that time. Back-mutations will throw the balance even further in the direction of selection. These and similar calculations lead to the conclusion that, indeed, natural selection, not mutation, drives evolution. Mutation does play the vital role of keeping temporarily unfit genotypes in the population at a low level – as a source for possibly useful variability when conditions change. Exercise 5 provides an introduction to the many theoretical studies concerning the balance between selection and mutation.

Discussion

From the genetic point of view, perhaps the most important achievement of this chapter is methodological – we have illustrated in a simple example the use of tools that have been extensively applied to the analysis of complex natural situations. But even the simple problem that we have investigated in detail provides interesting examples. Thus, the situation in which the heterozygote is most fit finds expression in Africa, where the gene whose homozygote is responsible for sickle-cell anemia confers some resistance to malaria.

Moreover, this situation of heterozygotic superiority provides a strong experimental confirmation of evolutionary theory in the demonstration by T. Dobzhansky on *Drosophila pseudo-obscura* that the approach to the *Aa* steady state is quantitatively as predicted by the equivalent of Figure 3.7 (Roughgarden, 1979, p. 41).

Our qualitative results have identified another possible force for biological variability in the opposite situation, where the heterozygote is least fit. We have seen that initial conditions determine which allele prevails. We would then expect to find both possibilities in different areas of the world, with different histories.

We have examined how randomly mating populations change when subject to a fixed selection regime on a single locus. For a better understanding of nature, we must also examine the effects of mutation, of assortative (nonrandom) mating, of multigenic traits, of density-dependent selection coefficients, of simultaneous selection on more than one trait, of dispersal in nonuniform environments, and so forth.

From the methodological point of view, most important is our finding of different qualitative behaviors for different values of the parameters K and L. Figure 3.5 sums up the results by means of a **parameter plane** that is divided up into several regions, each of which corresponds to a distinct behavior. In other circumstances there may be a **parameter space** of many dimensions. As we shall see later, notably in Chapters 6 and 9, the concept of a multidimensional parameter space divided up into regions of different qualitative behaviors may be of considerable help in understanding complex biological systems.

Exercises

1. (a) Verify equations (6) and (9). In particular, show that if $K+L=0$, then there are no solutions to (8) other than (9a) and (9b).
   (b) Verify equation (15).
   (c) Verify equation (19), taking advantage of the fact that a cancellable factor of $K+L$ appears in the evaluation of the denominator of (15).
   (d) Precisely why can we employ equation (16) to obtain (21a) for p_n when $0 < p_n \ll 1$?
   (e) Check that the approximating straight line in Figure 3.8 has the correct slope.
2. Show that the graphs of $L=K/(2K-1)$ and $L=K/(K-1)$ are correctly given in Figure 3.1 by verifying the asymptotes and the regions of increase and decrease, and checking the concavity.
3. Use a hand calculator, preferably a programmable one, to test the conjectured conclusions of Figure 3.5. Choose K and L (different from the values used in Figures 3.3 and 3.4) so that each of the four biologically meaningful regions in the K–L plane is represented. For each region, select at least two initial conditions and calculate a number of p_n values.

4. In an article on weed resistance to herbicides, Gressel and Segel (1978) considered (among other things) the following model for the number R_n per square meter in year n of a certain type of weed:

$$R_{n+1} = fR_n + \mu, \quad n = 0, 1, 2, \ldots.$$

Here μ is a (constant) mutation frequency, and f is the fitness ratio, a constant that quantifies the relative fitness of the given weed species and its principal competitor.

(a) If the steady-state solution is denoted by $R_n = A$, find A.

(b) Define T_n by $T_n = R_n - A$. Show that $T_{n+1} = fT_n$. Find the general solution of this equation. What is the corresponding formula for R_n if $R_0 = K$, with K a constant? Given that $f < 1$, interpret this result.

5. This problem concerns a special case of our discussion of population genetics, where $\Phi_{aa} = 0$ (i.e., $L = 1$).

(a) Show that in this case

$$p_{n+1} = \frac{1 - Kp_n}{2 - (1+K)p_n} \quad \text{if } p_n \neq 0, \quad p_{n+1} = 0 \quad \text{if } p_n = 0, \tag{29}$$

with steady states $p_n = p$ given by

$$p = 0, \quad p = \frac{1}{1+K}, \quad p = 1.$$

(b) Because of the unusual behavior of (29), the case $p = 0$ must be treated separately. But the other two cases can be handled as usual. Thus, show that if small perturbations r_n and q_n are defined by

$$p_n = \frac{1}{1+K} + r_n, \quad p_n = 1 + q_n,$$

then

$$r_{n+1} = (1-K)r_n, \quad q_{n+1} = \frac{1}{1-K}q_n.$$

(c) What can be concluded from part (b) concerning the stability of the steady states $p = 1/(1+K)$ and $p = 1$? Do these results make sense biologically?

(d) Analysis of the stability of $p = 0$ cannot be completed by mechanically carrying out our standard procedure; yet no fundamentally new ideas are required to do this. What, in fact, can be said about the stability of $p = 0$? Is your answer biologically reasonable?

(e) If mutations are considered in the case $\Phi_{aa} = 0$, the expression (1) for the total number of A alleles is taken to be

$$2\Phi_{AA}Np_n^2 + 2\Phi_{Aa}Np_nq_n - \mu(2\Phi_{AA}Np_n^2 + 2\Phi_{Aa}Np_nq_n)$$
$$+ \mu'(2\Phi_{Aa}Np_nq_n).$$

What has been assumed? Exactly what are μ and μ'? (We could now continue with an analysis of the effect of mutations, but we shall not do so here.)

6. Examine equation (6) when $K = 0$, $L \neq 0$ and when $L = 0$, $K \neq 0$. Discuss the biological appropriateness of the results.

7. The text's treatment ignored differences in fertility among genotypes. To take this into account, let surviving adults give rise respectively to

$2m_{AA}$, $2m_{Aa}$, and $2m_{aa}$ surviving gametes (equally divided between eggs and sperm). Define "absolute selective values" (an overall measure of fitness) by

$$W_{AA} \equiv m_{AA}\Phi_{AA}, \quad W_{Aa} \equiv m_{Aa}\Phi_{Aa}, \quad W_{aa} = m_{aa}\Phi_{aa}.$$

Show that there results an equation of the form (3), the only difference being that the W's replace the Φ's.

8. Take $\Phi_{AA} = \Phi_{Aa} = \Phi_{aa}$ in (3). Show that the result is to be expected on biological grounds.

4

Enzyme kinetics

Biochemical kinetics is an area in which the necessity for mathematical modeling has long been evident. Mechanisms for reactions are inferred by comparing different types of experimental graphs with various theories. Particularly subtle and interesting are the kinetics of enzyme actions, and this chapter will provide an introduction to that subject.

We begin with a review of the "law" of mass action, the accepted model for describing kinetics at sufficiently low concentrations of reactants. When this model is applied to the simplest case of enzyme action there results an analytically unsolvable system of equations. It is customary and useful to approximate this system by making the "quasi-steady-state" assumption that the enzyme-substrate complex is effectively at a steady-state concentration with respect to the current substrate concentration. We shall analyze this assumption in some detail.

The other major topic of this chapter is a theoretical analysis of several types of cooperative enzyme kinetics. We employ one of the major theories – that associated with the names of Monod, Wyman, and Changeux. A basis is provided for approaching the large literature concerning inference of molecular mechanisms from concentration measurements under various conditions. This is not our main purpose, however. Rather, we wish to lay a foundation for an exploration of the dynamic consequences of the nonlinear feedbacks that often occur in enzymatic systems (see Chapter 6).

Law of mass action

Suppose that a molecule A can combine with a molecule B to form a combined molecule C, and C in turn can break apart into its original constituents A and B. At time t, let the respective concentrations (number of molecules per unit volume) be denoted by $A(t)$, $B(t)$, and $C(t)$.

We make the reasonable and conventional assumption that the rate of formation of C is jointly proportional to the concentrations of A and B. According to this assumption, if the concentration of B (for example) is doubled or trebled, the formation rate of C is correspondingly doubled or trebled. Such an assumption is justified at sufficiently low concentrations, where each molecule moves independent of all other molecules. In this situation, increasing the number of B molecules will proportionately increase the number of A–B collisions that give rise to C.

We also assume that a given C molecule acts entirely independent of other C molecules in that it has a certain fixed probability per unit time of breaking apart into its constituents. This is equivalent to the assumption that the rate at which C molecules break apart is proportional to the number of such molecules that are present. For example, if 10^5 molecules were present in a given volume, 10 per second might break apart; if 10^6 were present, then 100 per second would split.

Combining our assumptions about the formation and disintegration of C, we introduce the proportionality constants k_1 and k_{-1} and write

$$dC/dt = k_1 AB - k_{-1} C. \tag{1}$$

When a molecule of C is formed, an A molecule and a B molecule "disappear," because they become part of C. Conversely, A and B molecules reappear when C breaks apart. Consequently,

$$dA/dt = -k_1 AB + k_{-1} C, \tag{2}$$

$$dB/dt = -k_1 AB + k_{-1} C. \tag{3}$$

The reasoning by which the differential equations (1), (2), and (3) were derived is associated with the term **law of mass action.** The proportionality factors k_1 and k_{-1} are called **rate constants**. The chemical reactions we have considered are conventionally summarized by the symbolism

$$A + B \underset{k_{-1}}{\overset{k_1}{\rightleftharpoons}} C. \tag{4}$$

If we wish to know how the concentrations $A(t)$, $B(t)$, and $C(t)$ behave in the course of time, we must know not only the rules by which the chemicals react but also how much chemical was present to start with. If we designate the starting time as $t = 0$, the **initial conditions** take the following form:

$$\text{At } t = 0, \quad A = A_0, \quad B = B_0, \quad C = C_0, \tag{5}$$

where A_0, B_0, and C_0 are given constants.

From the initial conditions (5) and the differential equations (1)–(3), we expect to be able to determine the functions $A(t)$, $B(t)$, and $C(t)$ for all times $t \geqslant 0$. We reserve detailed discussion for the next example, which is somewhat more complicated and interesting. Properties of the solution to (1), (2), (3), and (5) (with $C_0 = 0$) are developed in Exercise 6.

Enzyme and substrate

Enzymes catalyze the transformation of molecules into altered forms. The molecule that is to be transformed is called the **substrate** of the reaction, and the transformed molecule is called a **product**. Typically the enzyme acts to enhance the transformation process by binding with the substrate to form an enzyme-substrate **complex**. This later can break apart into the original enzyme plus the altered substrate (product). The desired transformation may fail to take place, in which case the disintegration of the complex yields the original enzyme and substrate. Assuming, for simplicity, that product formation can be regarded as an irreversible step, the reactions in question can be symbolized by

$$E + S \underset{k_{-1}}{\overset{k_1}{\rightleftarrows}} C \overset{k_2}{\longrightarrow} E + P. \tag{6}$$

Here, E, S, C, and P denote the enzyme, substrate, complex, and product, respectively, and we shall use the same letters to denote the concentrations of these quantities. Applying the law of mass action, we find that the following are the differential equations and initial conditions that correspond to (6):

$$dE/dt = -k_1 ES + (k_{-1} + k_2)C, \tag{7a}$$

$$dS/dt = -k_1 ES + k_{-1} C, \tag{7b}$$

$$dC/dt = k_1 ES - (k_{-1} + k_2)C, \tag{7c}$$

$$dP/dt = k_2 C, \tag{7d}$$

$$E(0) = E_0, \quad S(0) = S_0, \quad C(0) = C_0, \quad P(0) = P_0. \tag{8}$$

Most commonly studied is a special case of the initial conditions,

$$E(0) = E_0, \quad S(0) = S_0, \quad C(0) = 0, \quad P(0) = 0, \tag{9}$$

where it is assumed that the reaction is just about to start, so that no complex and no product are as yet present. [Note that the product concentration P can be found from (7d) once equations (7a–c) are solved for E, S, and C.]

Computers can be used to trace the evolution of the concentrations $E(t)$, $S(t)$, $C(t)$, and $P(t)$ for any given set of parameters k_1, k_{-1}, k_2,

E_0, S_0, C_0, and P_0. Nonetheless, a complete study of even the relatively simple equation set (7) and (8) would be difficult to carry out numerically, for presentation of all possible solutions would require books of graphs. Especially to obtain insight into the processes involved, analytic solutions are useful, but unfortunately they are beyond our powers even for the relatively simple problem under consideration. We thus present the so-called quasi-steady approximation, which provides a relatively simple and useful view of the enzyme-substrate interaction. Our treatment is more detailed than those generally given, in the sense that the reasons for expecting the approximation to work are more carefully discussed. We wish not only to justify a result that is very often used in biochemistry but also to illustrate how theoreticians go about their craft.

Before beginning our discussion of approximations, let us take note of one general result. If we add equations (7a) and (7c), we find that

$$dE/dt + dC/dt = 0 \tag{10}$$

or

$$d(E+C)/dt = 0, \quad \text{i.e., } E(t) + C(t) = \text{constant.}$$

The constant can be evaluated from the initial conditions (8):

$$E(t) + C(t) = E_0 + C_0. \tag{11}$$

Intuitively this result follows at once from the fact that enzyme molecules are always to be found in either free (E) or complexed (C) form. (No degradation or reconstitution of enzyme has been taken into account.)

We shall use (11) to eliminate the enzyme concentration E from our equations. Writing

$$E(t) = E_0 + C_0 - C(t), \tag{12}$$

we see that (7b) and (7c) become

♦♦ $$dS/dt = -k_1(E_0 + C_0 - C)S + k_{-1}C, \tag{13a}$$

♦♦ $$dC/dt = k_1(E_0 + C_0 - C)S - (k_{-1} + k_2)C. \tag{13b}$$

Together with the initial conditions

♦♦ $$S(0) = S_0, \quad C(0) = C_0, \tag{14}$$

(13a) and (13b) constitute equations for two unknown functions $S(t)$ and $C(t)$. If these equations could be solved, the concentration $P(t)$ could be separately determined, from (7d), whereas, of course, $E(t)$ is given by (12).

The mathematical problem composed of the differential equations (13a) and (13b) and the initial conditions (14) is typical in that it is apparently insoluble, in the sense of finding some manageable formula for the solution under all circumstances. In such cases, often we can make progress by limiting ourselves to certain special sets of parameter choices for which considerable simplifications are possible.

In the present instance, it is often the case that the initial amount of substrate present is "large" in the sense that the substrate concentration remains virtually unchanged for quite a considerable period of time. To gain understanding of these situations, let us first make the somewhat radical assumption that by some means, S is permanently held at its initial value, that is, $S(t) \equiv S_0$ [so that (13a) is no longer relevant]. Then (13b) becomes

$$dC/dt = k_1(E_0 + C_0 - C)S_0 - (k_{-1} + k_2)C. \tag{15}$$

This equation has the form

$$dC/dt = (\text{constant}) - (\text{constant} \cdot C)$$

and is thus quite easy to solve (see the supplement at the end of this chapter.) Remembering the initial condition

$$C(0) = C_0, \tag{16}$$

we find

$$C(t) = (C_0 - \bar{C})\exp[-(k_1 S_0 + k_{-1} + k_2)t] + \bar{C}, \tag{17}$$

where we have made the abbreviations

$$\bar{C} \equiv (E_0 + C_0)S_0/(K_m + S_0), \quad K_m \equiv (k_{-1} + k_2)/k_1. \tag{18a,b}$$

Time scale

As is indicated in Figure 4.1, the solution (17) departs from its initial value C_0 and approaches closer and closer to an asymptote $\bar{C}$. Although completion of the shift from C_0 to $\bar{C}$ takes infinitely long, it is important for future considerations to have an idea of *how long it takes for a significant part of the change in the function $C(t)$ to occur*. This is called the **time scale** of the change. If (17) is written in the form

$$C(t) = C_0 e^{-kt} + \bar{C}(1 - e^{-kt}), \quad k \equiv k_1 S_0 + k_{-1} + k_2, \tag{19a,b}$$

it becomes evident that the time scale is the time it takes for e^{-kt} to change significantly, for in that time $C_0 e^{-kt}$ will decrease significantly

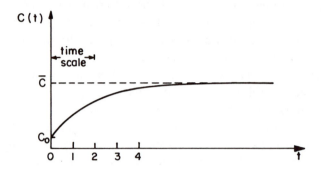

Figure 4.1. Graph of equation (17), or (19), which is the same function. In the notation of (19), k has been taken to be 0.5. The time scale $k^{-1}=2$ is shown.

from its initial value C_0, while $\bar{C}(1-e^{-kt})$ will increase significantly from its initial value zero toward its final value $\bar{C}$. *The time scale for the function e^{-kt} is generally taken to be k^{-1},* for in time k^{-1} this function decreases by a factor of e^{-1}, that is, because $e \approx 2.7$, to about one-third of its original value. To see this, note that if $f(t)=\exp(-kt)$, then

$$f(t+k^{-1})=e^{-k(t+k^{-1})}=e^{-kt}e^{-1}=e^{-1}f(t). \tag{20}$$

That is,

$$f(t+k^{-1}) \approx \tfrac{1}{3}f(t).$$

Bear in mind that the definition of time scale is not precise: A decrease in e^{-kt} by a factor of $e^{-2}=0.14$ takes place in a time of approximately $2k^{-1}$, but if $k^{-1}=0.6$ sec, say, then if we take a time scale of k^{-1} or $2k^{-1}=1.2$ sec or $3k^{-1}=1.8$ sec, we obtain the same result for present purposes – that $C(t)$ changes appreciably in a time of the order of 1 second.

A characterization of the time scale that allows its ready calculation is an estimate of how long it takes for the function to change from its maximum to its minimum value, assuming that the function is changing at its maximum rate. That is:

$$\text{Time scale of } f(t) \approx \frac{f_{\max}-f_{\min}}{|df/dt|_{\max}}. \tag{21}$$

For the function $f(t)=\exp(-kt)$ for $0 \leqslant t < \infty$,

$$f_{\max}=1, \quad f_{\min}=0, \quad |df/dt|_{\max}=|-ke^{-kt}|_{\max}=k,$$

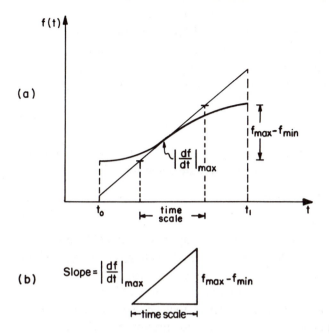

Figure 4.2. Aspects of definition (21) for the time scale. (a) The ingredients needed to define the time scale. (b) The time-scale triangle; details from (a).

once again giving a time scale of k^{-1}.[1] Figure 4.2 shows that the characterization (21) provides a reasonable estimate for the time it takes for a significant change in the function to take place. For further discussion of scales and related matters, see Lin and Segel (1974, Section 6.3).

The quasi-steady state

Let us return to the equations (13) and the initial conditions (14) for the substrate concentration $S(t)$ and the complex concentration $C(t)$. We recall that if S can be considered a constant, S_0, then C approaches an asymptote $\bar{C}$, regardless of its initial condition. As the asymptote is approached, $C(t)$ changes more and more slowly, so that it is no surprise that $\bar{C}$ satisfies the steady-state equation obtained by setting $dC/dt = 0$ in (13b). (This is shown in the supplement.) We now wish to consider what

---

1 Strictly speaking, f has no minimum for $0 \leqslant t < \infty$. The number zero is the *greatest lower bound* or *infimum* (inf) [i.e., the largest number d satisfying $f(t) > d$ for $0 \leqslant t < \infty$]. For full accuracy, (21) should be phrased in terms of inf and its "upper" counterpart sup, but most readers will not wish to dwell on this refinement.

will happen if S is no longer a constant, but changes gradually. In such circumstances we assert that, to a good approximation, at any time, C will approximately equal the steady-state value appropriate to the instantaneous concentration $S(t)$. To be more precise, if $S = S_0 = $ constant, then (17) and (18a) show that $C(t)$ approaches a steady state given by

$$\bar{C} = \frac{(E_0 + C_0)S_0}{K_m + S_0}. \tag{22}$$

We assert that if S changes "slowly enough," C will, to a good approximation, be given by the **quasi-steady-state equation**

◆◆
$$C(t) = \frac{(E_0 + C_0)S(t)}{K_m + S(t)}. \tag{23}$$

How slowly is "slowly enough"? The approximation (23) should be a good one if the time scale for the change of S is large compared with the time scale that characterizes the change in C, for then $C(t)$ remains almost fully adjusted to the instantaneous value of $S(t)$.

An analogy may be helpful. Suppose that an airplane C is trying to maintain a fixed position with respect to an airplane S. If S is flying with constant velocity, then after an initial transient period, C can take up its desired position. Even if S tries to take evasive action, C can maintain its position with respect to S – to a good approximation – providing that C is faster and more maneuverable than S.

To put the matter slightly differently, suppose that S is continually changing its velocity, but the time scale for C to adjust its course is much shorter than the corresponding time scale for S. Then C will be moving unsteadily with respect to a fixed reference frame (such as the ground), but C can be said to be in a quasi-steady state, for C steadily maintains its position relative to S (Figure 4.3).

It should now not puzzle the reader that (23) is obtained from (13b) by setting $dC/dt = 0$ and solving for the complex concentration C in terms of the substrate concentration S. That is, the *quasi-steady-state hypothesis for C assumes that $dC/dt \approx 0$* even though C is not in fact a constant. All that is required is that S change slowly enough, in the sense that we have stated, so that C behaves as if S were "temporarily" constant.

Consequences of the quasi-steady-state approximation

According to the quasi-steady-state approximation, the complex concentration C is given in terms of the substrate concentration S by (23).

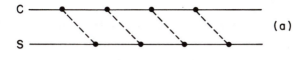

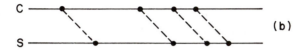

Figure 4.3. Relationships between airplanes C and S, where C is attempting to maintain a position a fixed distance above and behind S. The points represent the positions of the planes at four successive minutes. (a) S and C both moving steadily in a straight line. (b) S tries to evade C by accelerating and decelerating, but C's time scale for changing speed is much smaller than that of S, so that C can maintain its position with respect to S.

Substituting for C in equation (13a) for dS/dt, we find, after a little algebra, the following equation for dS/dt in terms of S:

$$dS/dt = -k_2(E_0 + C_0)S/(K_m + S). \tag{24}$$

Also, equation (7d) implies that the product concentration P can be found from

$$dP/dt = k_2(E_0 + C_0)S/(K_m + S). \tag{25}$$

We see from (24) and (25) that according to the present approximation the rate of substrate disappearance is equal to the rate of product formation. Such a balance must hold if the complex concentration is to remain "in equilibrium" with the present substrate concentration.

An explicit solution of equation (24) for S is discussed in Exercise 3. Rather than compare experiments with this solution, it is more common to compare measurements of dS/dt with the predictions of the equation. Let us turn to this matter.

The common value of $|dS/dt|$ and dP/dt is called the **velocity of the reaction** and is denoted by V. From now on we shall restrict ourselves to the usual case in which $C_0 = 0$. In addition, we take account of the fact that the velocity usually is measured before S decreases appreciably from its initial value S_0. Putting $C_0 = 0$, $S = S_0$ in (24) and (25), we can thus write

$$V = k_2 E_0 S_0/(K_m + S_0). \tag{26}$$

(V should now be termed the **initial velocity**, but in accord with convention we generally shall not use this refined terminology.) Because (Exercise 1)

$$\partial V / \partial S_0 > 0, \tag{27}$$

the reaction velocity is an increasing function of the substrate concentration S_0 (as is to be expected). Thus, the maximum value of V, V_{max}, is approached for very large values of S_0. In mathematical terms,

$$V_{max} = \lim_{S_0 \to \infty} V = k_2 E_0. \tag{28}$$

In terms of V_{max}, equation (26) can be written

◆◆ $$V = \frac{V_{max} S_0}{K_m + S_0}. \tag{29}$$

The **Michaelis constant** K_m (as it is called) is now seen to give the initial substrate concentration at which the reaction velocity is half maximal, for if $S_0 = K_m$ is substituted into (29), we obtain $V = \frac{1}{2} V_{max}$.

V can be measured as a function of S_0, enabling the maximum velocity V_{max} and the Michaelis constant K_m to be obtained by curve fitting. It is convenient to fit data to a straight line. With this in mind, let us rearrange (29) as follows:

$$\frac{1}{V} = \frac{1}{V_{max}} \left(1 + \frac{K_m}{S_0} \right). \tag{30}$$

In terms of the variables

$$y = 1/V, \quad x = 1/S_0, \quad m = K_m / V_{max}, \quad b = 1/V_{max},$$

equation (30) becomes

$$y = mx + b. \tag{31}$$

In this familiar notation, $1/V_{max}$ is immediately recognizable as the "y intercept" (where $x = 0$), and K_m can be determined from the slope K_m / V_{max}. Alternatively, it can be recognized that when $y = 0$, $x = -1/K_m$.

The straight-line graph that comes from the rearrangement (30) of equation (29) is called a **Lineweaver-Burk plot** (Figure 4.4).

Validity of the quasi-steady-state approximation

The quasi-steady-state approximation is used a great deal; in any given experimental situation it is important to be able to tell whether or not the approximation is valid. In order to find an explicit condition, we shall make quantitative the implicit requirement that the time scale for substrate changes be long compared with the time scale for complex

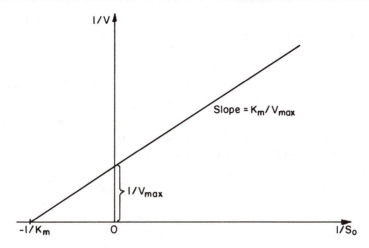

Figure 4.4. The Lineweaver–Burk plot of relation (30) between the reaction velocity V and the initial substrate concentration S_0.

changes. The latter scale has already been estimated as k^{-1}. To estimate the former, we employ equation (21). Because S changes from a maximum of S_0 to a minimum of zero, and because $|dS/dt|$ is maximal when $S = S_0$ [by (27)], we see from (24), with $C_0 = 0$, that

$$\frac{\text{time scale for}}{\text{changes in } S(t)} = \frac{S_0}{k_2 E_0 S_0/(K_m + S_0)} = \frac{K_m + S_0}{k_2 E_0}. \tag{32}$$

The condition we seek is that this scale be large compared with the time scale for changes in $C(t)$, k^{-1}. That is,

$$k^{-1} \ll (K_m + S_0)/k_2 E_0. \tag{33a}$$

Inserting expressions (19b) and (18b) for k and K_m, we obtain as the final form of the condition

$$k_2 E_0/k_1 \ll (K_m + S_0)^2. \tag{33b}$$

Equation (33b) will certainly hold if the initial substrate concentration S_0 is sufficiently large compared with the initial enzyme concentration E_0. That the quasi-steady-state approximation can be used when $S_0 \gg E_0$ was pointed out by Briggs and Haldane (1925). The approximation will also be valid if the product formation rate k_2 is sufficiently small, an approach associated with Michaelis and Menten (1913). But the Briggs–Haldane and Michaelis–Menten conditions are just two important special cases of the general relation (33b) for justifying the use of quasi-steady-state approximation results such as (23) and (26).

In general, we would like to be able to know in advance when a quasi-steady-state assumption might be used to simplify a system of equations. This is not a simple matter, for it requires guessing the time scale of an unknown solution. One recommended procedure is to introduce the quasi-steady-state assumption where intuition indicates that it might be appropriate, solve the resulting simplified problem, and then check back for consistency. More hints on this subject have been provided by Lin and Segel (1974, Chapter 6). (Without further experience, readers of this book should not expect to be able to forecast reliably when a quasi-steady-state assumption can be introduced into a new situation – but it is hoped that they now understand the issues involved.)

Cooperative kinetics

The straight line predicted by the Lineweaver–Burk expression (30) for the reciprocal reaction velocity has been observed in a number of instances, but in many other cases measurements have shown marked deviation from the predicted graph. Theoretical explanations for the observed results are the "allosteric" theory of Monod, Wyman, and Changeux (1965) and the related "induced-fit" theory of Koshland, Nemethy, and Filmer (1966). We shall now present some salient features of the MWC theory of Monod et al. (1965).

Two basic hypotheses will be made: first, that the enzyme is an **oligomer** composed of several identical subunits; second, that the subunits exist in two conformations. The two different shapes are assumed to be associated with different reaction rates. According to the MWC theory, transitions between the conformations are **concerted**, in that any transition is made simultaneously by all subunits of the oligomer. We shall here consider the simplest case of two subunits (a **dimer**). As a simplification, instead of considering the general case of a more reactive and a less reactive conformation, we shall confine ourselves to the situation in which one of the conformations is entirely unreactive.

Not only enzyme molecules but also other proteins are found to be oligomers with multiple conformational states. For the moment, let us consider the more general situation in which a protein binds some smaller molecule to its subunits. The latter molecule is often called a **ligand** (Latin: *ligare* – to bind); the binding forms a protein-ligand complex.

In accord with the simplest version of the MWC theory, we shall assume that each subunit of the oligomer can bind ligand at a single site. Moreover, we assume that the transition between conformations (with

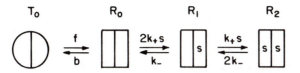

Figure 4.5. Diagram illustrating the states of an MWC dimer with exclusive binding to the R state.

forward and backward rates f and b) can occur only when the protein molecule is completely free of ligand. Thus, the possible conformations of the postulated dimer are as shown in Figure 4.5. In accord with common practice, we employ the notation T for the concentration of protein in the "tight" inactive state and R for the reactive state. A subscript will designate the number of bound ligand molecules. We denote the ligand concentration by s.

Ligand is assumed to bind to a given subunit with rate constant k_+ and to dissociate with rate constant k_-. Then the concentrations of the various states are governed by the following differential equations:

$$\frac{dT_0}{dt} = -fT_0 + bR_0, \tag{34a}$$

$$\frac{dR_0}{dt} = fT_0 - bR_0 + k_- R_1 - 2k_+ R_0 s, \tag{34b}$$

$$\frac{dR_1}{dt} = -k_- R_1 + 2k_+ R_0 s + 2k_- R_2 - k_+ R_1 s, \tag{34c}$$

$$\frac{dR_2}{dt} = -2k_- R_2 + k_+ R_1 s, \tag{34d}$$

$$\frac{ds}{dt} = -2k_+ R_0 s + k_- R_1 + 2k_- R_2 - k_+ R_1 s. \tag{34e}$$

(Initial conditions are needed to specify the problem fully.)

The appearance of the so-called **statistical factor** 2 deserves comment. In the last term of (34b), for example, this factor appears because there are two sites on the free R dimer to which a ligand can bind. By definition, $k_+ R_0$ gives the number of binding events per unit time if R_0 sites per unit volume are available for binding. But because R_0 denotes the concentration of free *dimer,* there are $2R_0$ sites available for binding, per unit volume.

Note that

$$0 = \frac{dT_0}{dt} + \frac{dR_0}{dt} + \frac{dR_1}{dt} + \frac{dR_2}{dt} = \frac{d}{dt}(T_0 + R_0 + R_1 + R_2) \tag{35}$$

or

$$T_0 + R_0 + R_1 + R_2 = \text{constant} \equiv E, \tag{36}$$

where E is the (unchanging) total number of protein molecules in a unit volume.

Percentage ligand bound in steady state

Operation of the postulated system is well characterized by the fraction Y of subunits to which a ligand is bound when the steady state is attained. Let us therefore derive an expression for Y as a function of the ligand concentration s. To do this, we must determine the steady-state values of the various configurations (denoted by a bar) from the equations

$$0 = -f\bar{T}_0 + b\bar{R}_0, \tag{37a}$$

$$0 = f\bar{T}_0 - b\bar{R}_0 + k_-\bar{R}_1 - 2k_+\bar{R}_0 s, \tag{37b}$$

$$0 = -k_-\bar{R}_1 + 2k_+\bar{R}_0 s + 2k_-\bar{R}_2 - k_+\bar{R}_1 s, \tag{37c}$$

$$0 = -2k_-\bar{R}_2 + k_+\bar{R}_1 s. \tag{37d}$$

Taking into account the fact that state R_1 has a single bound site and that R_2 has two, the desired expression for Y can then be obtained from

$$Y = \frac{\bar{R}_1 + 2\bar{R}_2}{2E}. \tag{38}$$

We have used the fact that the total number of sites, free and bound, is $2E$, and s has been regarded as constant.

Solving (37d), we obtain

$$\bar{R}_2 = s\bar{R}_1/2K, \tag{39}$$

where

$$K \equiv k_-/k_+ \tag{40}$$

is the **dissociation constant** of the binding reaction.

If we add (37d) and (37c), we obtain

$$-k_{-1}\bar{R}_1 + 2k_+\bar{R}_0 s = 0, \tag{41}$$

which yields

$$\bar{R}_1 = 2s\bar{R}_0/K. \tag{42}$$

Equations (41) and (42) have been derived very simply from the governing equations, but there is a way of looking at the problem that makes the derivation even faster – once one has become used to it. This way stems from the observation that at steady state there must be a "detailed balance" between the transition $R_1 \to R_2$ and the inverse transition (or back-reaction) $R_2 \to R_1$. Because there is no net flow to R_1 from the reaction $R_1 \rightleftarrows R_2$, we can ignore this reaction in considering the steady-state value of R_1. "Operationally," this can be accomplished by covering with one's hand the part of Figure 4.5 that is to the right of R_1. This leads to the balance condition (41) as a condition for the steady state. We now know that there is no net flow to R_0 from R_1, so that we can obtain the next balance equation by covering all of Figure 4.5 that is to the right of R_0. The resulting equation is (37a), which demonstrates again that the "covering" method we have described is just a quick way of finding what in any case can be deduced directly from the governing equations.

Whether obtained directly from (37a) or by use of the covering method, the final equation that we need is $-f\bar{T}_0 + b\bar{R}_0 = 0$, or

$$\bar{R}_0 = \bar{T}_0/L, \tag{43}$$

where the **allosteric constant** L is defined by

$$L \equiv b/f. \tag{44}$$

Combining (38), (39), (42), and (43), we easily find (Exercise 4) that

◆◆
$$Y = \frac{S^2 + S}{L + (1+S)^2}, \quad \text{where } S \equiv \frac{s}{K}. \tag{45a,b}$$

It is normally assumed that $L \gg 1$, so that in the absence of ligand almost all the dimers will be in the inactive T state.

For the special case $L = 0$, the T state never occurs, and (45) reduces to

$$Y = \frac{S}{1+S}. \tag{46}$$

[Note the elegant parameter-free form (46), a consequence of using the dimensionless variable S.]

As a check, let us derive (46) directly. When $L = 0$, we have $2E$ identical sites, of which $2EY$ are bound and $2E(1-Y)$ are free. At steady state, the rate at which bound sites dissociate must equal the rate at which free sites bind a ligand. Thus,

$$2EYk_- = 2E(1-Y)k_+ s,$$

which leads at once to (46).

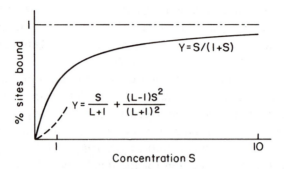

Figure 4.6. Fraction of sites bound, Y, for noncooperative binding (solid line) and for the MWC dimer at low ligand concentration (dashed line). The concavity of the latter curve is a sign of positive cooperativity.

Let us analyze (46) as a standard against which to measure more complicated binding functions. We note that

$$\frac{\partial Y}{\partial S} = (1+S)^{-2}, \quad \frac{\partial^2 Y}{\partial S^2} = -2(1+S)^{-3}. \tag{47a,b}$$

The positivity of $\partial Y/\partial S$ is due to the fact that there is more binding at higher ligand concentration. For small S, $Y \approx S$. This reflects a constant probability that each new ligand is bound. The increase in binding slows at higher concentrations according to (47b) $[\partial(\partial Y/\partial S)/\partial S < 0]$, because fewer and fewer free sites are available. There results a characteristic **saturated** graph (Figure 4.6).

Expression (45) is somewhat complicated to analyze in full. It will prove sufficient to restrict ourselves to small values of S, for which [Exercise 4(b)] we have the Taylor approximation

$$Y \approx \frac{1}{L+1} S + \frac{L-1}{(L+1)^2} S^2. \tag{48}$$

If S is very small, $Y \sim S$, and binding is proportional to ligand concentration as before. But the full expression (48) shows that binding *increases* as ligand concentration increases (remember that $L \gg 1$), in marked contrast to (47) (Figure 4.6). Such an increase is a sign of what is termed **positive cooperativity**, a situation in which each ligand that binds makes easier the binding of its successor. In the present instance, the cooperativity comes about in the following way. At a very low value of S, almost all the subunits are in the unreactive state T_0. Only a very few subunits are in the state R_0 that will bind ligand. As ligand concentration increases, however, an appreciable number of R_1 subunits will be found,

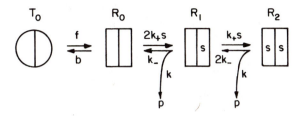

Figure 4.7. Modification of Figure 4.5 to allow for product formation.

each with an empty site that is available for binding. Of course, at still higher ligand concentrations [beyond the validity of (48)], only a small number of dimers are in state T_0, and small effects of positive cooperativity are overridden by the usual effects of saturation (see Exercise 8).

The cooperative enzymatic dimer

Let us return to the case in which the protein is an enzyme. Now the ligand is a substrate for the reaction (so that the letter s becomes more appropriate to denote its concentration). As before, there are binding and dissociation reactions of s with each subunit. In addition, we shall take into account an irreversible reaction, at rate k, in which a complex breaks apart into the original subunit plus a product molecule (concentration p). Figure 4.5 must now be supplemented: For every possibility of dissociation at rate k_-, we must add the possibility of product formation at rate k (Figure 4.7). Wherever k_- appeared in the kinetic equations (34), we must now write $k_- + k$. Moreover, there is a product-formation equation

$$dp/dt = 2kR_2 + kR_1. \tag{49}$$

We can rearrange (49) in the form

$$\frac{dp}{dt} = k\,\frac{2R_2 + R_1}{2E}\,2E \equiv 2kYE. \tag{50}$$

In (50), $Y(t)$ gives the percentage of bound sites. The total number of sites (per unit volume) is $2E$, so that $2EY$ is the total number of bound sites. To find the rate of product formation, we multiply this number by k, the rate of product formation per bound site – and this indeed is the product-formation equation (50).

We now make a generalization of our earlier quasi-steady-state assumption and assume that all the enzymatic forms are in a quasi-steady state. This allows us to employ the true steady-state result (45) for Y, and at once we have a formula for the rate of product formation:

$$\frac{dp}{dt} = 2kE \frac{S_m^2 + S_m}{L + (1 + S_m)^2}. \tag{51}$$

Here, because k_- must be replaced by $k_- + k$,

$$S_m(t) = \frac{s(t)}{K_m}, \quad K_m = \frac{k_- + k}{k_+}. \tag{52}$$

If V is used to denote the reaction velocity dp/dt and V_{max} is employed as an abbreviation for $2kE$ [see Exercise 4(c)], then (51) can be written

◆◆
$$\frac{V}{V_{max}} = \frac{S_m^2 + S_m}{L + (1 + S_m)^2}. \tag{53}$$

This formula is entirely analogous to (45a), and our remarks concerning cooperativity are immediately applicable.

Product regulation

Enzymes are of immense importance in biochemistry because of their catalytic function. Almost more important, however, is their ability to modify their catalytic efficiency and therefore to control the myriad interlocking reactions that are found in living matter. We have just examined an example of nontrivial substrate regulation, wherein the rate of reaction can depend strongly on the amount of substrate present. For later use, and as another example of what can occur, we now present a relatively simple case in which a second modifier molecule affects the rate of catalysis.

As before, we shall assume an MWC model with exclusive binding to the R state. In addition to a ligand s, a second molecule p is assumed to bind to the enzyme and also to "lock" it in the reactive R state – so that, once again, only free R can make the transition to T. Figure 4.5 now is generalized to Figure 4.8, with the binding and dissociation constants for p being denoted by k_+^* and k_-^*. The notation R_{ij} indicates that i molecules of s and j molecules of p are bound to the dimer $(i, j = 0, 1, 2)$.

Figure 4.8 is easily analyzed by the covering method described earlier. Indeed, it is assumed that transitions such as $R_{01} \leftrightarrows R_{11}$ do not occur in order to permit the use of this simple method. [It is possible to show that consideration of these transitions would not alter the results. Most of the basic ideas required are presented in Section 2.6 of Rubinow (1975).]

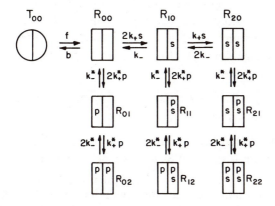

Figure 4.8. Generalization of Figure 4.5 to allow for regulation by a second molecule p.

As the reader should verify [Exercise 5(a)], the following equations describe the steady-state concentrations of the various constituents depicted in Figure 4.8:

$$R_{j2} = \frac{pR_{j1}}{2K_p}, \quad R_{j1} = \frac{2pR_{j0}}{K_p}, \tag{54}$$

$$R_{20} = \frac{sR_{10}}{2K_s}, \quad R_{10} = \frac{2sR_{00}}{K_s}, \quad R_{00} = \frac{T_{00}}{L}. \tag{55}$$

Here, K_s and K_p denote the dissociation constants for the s and p bindings:

$$K_s = \frac{k_-}{k_+}, \quad K_p = \frac{k_-^*}{k_+^*}. \tag{56}$$

The conservation relation (36) is here generalized to

$$T_{00} + \sum_{i,j=0}^{2} R_{ij} = E. \tag{57}$$

Let us introduce the dimensionless concentrations

$$S = \frac{s}{K_s}, \quad P = \frac{p}{K_p}. \tag{58}$$

Then the percentage of subunits to which S is bound is given by [Exercise 5(a)].

$$Y = \frac{(S^2 + S)(1 + P)^2}{L + (1 + S)^2 (1 + P)^2}. \tag{59}$$

For later use, we must generalize our results to the case in which p is a product obtained irreversibly from s with rate constant k. We assume that the enzymatic reactions are all in a quasi-steady state. Proceeding exactly as in the passage from (45a) to (51), we find that the rate of product formation is given by

$$\frac{dp}{dt} = \frac{2kE(S_m^2 + S_m)(1+P)^2}{L + (1+S_m)^2(1+P)^2}. \tag{60}$$

Here, $P(t)$ is still defined as in (58), but now in the definition of $S_m(t)$ the dissociation constant is replaced by the Michaelis constant K_m, as in (52).

Equation (60) is an example of reaction regulation by both substrate and product. Subtle and important effects of such feedbacks will be explored in Chapter 6.

Exercises

1. Verify equations (24) and (27).
2. The following explicitly solvable example illustrates several of the matters involved in the quasi-steady-state approximation. The reader is asked to fill in the details.
   (a) Consider the problem

   $$\frac{dx}{dt} = -2x+y, \quad \frac{dy}{dt} = -100(y-x), \quad y(0)=0, \quad x(0)=1. \tag{61a,b,c,d}$$

   Because of the factor 100, it appears that dy/dt is generally much larger than dx/dt. We thus assume that y is in a quasi-steady state with respect to x [i.e., from (61b), that $y \approx x$]. Using this to eliminate y in (61a), show that our approximation is

   $$x \approx \exp(-t), \quad y \approx \exp(-t).$$

   (b) Solve the differential equations without approximation (except for final numerical evaluation of exponents), thereby showing that for arbitrary constants A and B,

   $$y = Ae^{-101t} + Be^{-t}, \quad x = -\frac{1}{100}Ae^{-101t} + \frac{99}{100}Be^{-t}.$$

   After imposition of initial conditions, verify the "exact" answer

   $$x = \frac{99}{100}e^{-t} + \frac{1}{100}e^{-101t}, \quad y = e^{-t} - e^{-101t}.$$

   Compare this with the approximate answer, paying particular attention to the behavior of y when t is small.
3. (a) Verify, by differentiation, that (24) and the initial condition $S(0) = S_0$ have the solution

   $$S + K_m \ln(S/S_0) = S_0 - k_2(E_0 + C_0)t. \tag{62}$$

(b) If you know how to handle separable equations, solve (24) and obtain (62).

4. (a) Verify equations (45) and (47).
   (b) Verify equation (48).
   (c) Show that in equation (51), dp/dt increases monotonically toward $2kE$ as $S \to \infty$.

5. (a) Verify equations (54), (55), and (59).
   (b) Write an equation for dp/dt and use it to derive (60).

6. Consider the chemical reaction of equation (4) with the initial conditions

   $$A(0) = A_0, \quad B(0) = A_0, \quad C(0) = 0.$$

   (a) Prove that $A(t) + C(t) = A_0$. Why is this obvious chemically?
   (b) Employing $B(t) + C(t) = A_0$, derive

   $$\frac{dC}{dt} = k_1(A_0 - C)^2 - k_{-1}C.$$

   Introduce dimensionless variables

   $$Q = C/A_0, \quad \tau = k_{-1}t$$

   and show that

   $$\frac{dQ}{d\tau} = \frac{(1-Q)^2}{\alpha} - Q$$

   for a certain parameter α.
   (c) Show that the equation for Q has two steady states, the smaller of which is given by

   $$\bar{Q}_1 = 1 + \frac{\alpha}{2} - \frac{1}{2}\sqrt{\alpha^2 + 4\alpha}\ .$$

   (d) Show that a small perturbation Q' to a steady state $\bar{Q}$ satisfies

   $$\frac{dQ'}{d\tau} = -\left(\frac{2(1-\bar{Q})}{\alpha} + 1\right)Q'.$$

   (e) Show that $1 - \bar{Q}_1 > 0$, and therefore that $\bar{Q}_1$ is a stable steady state. Remembering that $Q(0) = 0$, use part (d) to conjecture the graph of Q as a function of t.
   (f) If you know how to solve separable equations, solve for $Q(\tau)$, and use the result to check the conjecture in part (e).

7. Show that when $S > 0$,

   $$\frac{S^2 + S}{L + (1+S)^2} < \frac{S}{1+S}.$$

   Justify this result in terms of (45) and (46).

8. Set $S^{-1} = x$ and use (A2.5) to show that (45) gives $Y \approx 1 - S^{-1} - (L-1)S^{-2}$. Thus demonstrate that for large S the graph is convex. Given this information, sketch the complete graph of (45) – recalling Figure 4.6. Thereby demonstrate the "sigmoid" shape that is associated with positive cooperativity (but see Exercise 10).

9. Consider equation (45) in Lineweaver-Burk form, wherein $1/Y$ is regarded as a function of $w = 1/S$. Show that

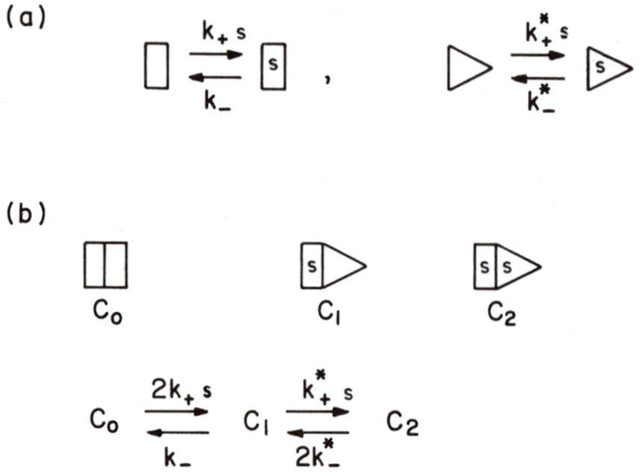

(a)

(b)

Figure 4.9. (a) Binding to subunits of two different configurations. (b) Binding to dimers.

$$\frac{d}{dw}\left(\frac{1}{Y}\right)=1+\frac{Lw(2+w)}{(w+1)^2}, \quad \frac{d^2}{dw^2}\left(\frac{1}{Y}\right)=\frac{2L}{(w+1)^3},$$

and therefore, in particular, that the graph is convex.

10. Consider a "free" dimer (with no ligand bound). Suppose that when a ligand (concentration s) binds to one subunit of the dimer, then the second subunit changes its conformation and thereby renders the next binding easier (positive cooperativity) or harder (negative cooperativity). Let C_i be the concentration of complex with i ligands bound ($i=0,1,2$). Let k_+ and k_- denote the binding and dissociation constants to a subunit of a free dimer (a dimer without bound substrate), with k_+^* and k_-^* denoting the corresponding quantities for a subunit with changed conformation (Figure 4.9).

(a) Show that at equilibrium (steady state), the percentage Y of bound sites satisfies

$$Y=\frac{K_a s+K_a K_a^* s^2}{1+2K_a s+K_a K_a^* s^2},$$

where the association constants K_a and K_a^* are given by

$$K_a=k_+/k_-, \quad K_a^*=k_+^*/k_-^*.$$

(b) Introduce the (dimensionless) quantities $z\equiv K_a s$, $\beta=K_a^*/K_a$ and obtain

$$Y=\frac{z+\beta z^2}{1+2z+\beta z^2}. \tag{63}$$

Show that

$$\frac{dY}{dz}=\frac{1+2\beta z+\beta z^2}{(1+2z+\beta z^2)^2}, \quad \frac{d^2Y}{dz^2}=2\frac{\beta-2-\beta z(3+3\beta z+\beta z^2)}{(1+2z+\beta z^2)^3}.$$

† (c) Show that $d^2Y/dz^2<0$ when $\beta<2$ and thus that there is no concave portion (and no sigmoidality) for this range of β. Nonetheless, we would

certainly regard positive cooperativity to be present if the association constant for the second binding is larger than that of the first (i.e., if $\beta > 1$). [For a "proof," see Section 1.4 of Segel (1980).] For $1 < \beta < 2$, we therefore have positive cooperativity and yet no region of concavity. Comment.

11. (a) Introduce $w = 1/z$ into (63) to obtain the Lineweaver–Burk form

$$\frac{1}{Y} = \frac{\beta + 2w + w^2}{\beta + w}.$$

(b) Show that $1/Y$ is an increasing function of w.

(c) Show that

$$\frac{d^2(1/Y)}{dw^2} = \frac{2\beta(\beta - 1)}{(\beta + w)^3}.$$

(d) From the definition of β it follows that the case $\beta < 1$ can appropriately be called an example of negative cooperativity. Why? (Thus, in this example at least, the type of cooperativity is made evident by the concavity or convexity of the Lineweaver–Burk plot.)

12. Generalize the basic MWC model to the situation in which the T state can also bind ligand, but more weakly than the R state. To Figure 4.5 add the transitions $T_0 \leftrightarrows T_1$, $T_1 \leftrightarrows T_2$. As a generalization of (45a), derive

$$Y = \frac{LcS(1 + cS) + S(1 + S)}{L(1 + cS)^2 + (1 + S)^2},$$

where $c = K/K_T$, K_T the dissociation constant for binding to a T monomer.

13. For the model of Exercise 12, show that if conversion to product occurs at different rates from the R and T states, then (53) generalizes to an equation of the form

$$V = \frac{V_T LcS_m(1 + cS_m) + V_S S_m(1 + S_m)}{L(1 + cS_m)^2 + (1 + S_m)^2}.$$

14. (a) Generalize (45a) to the case of a tetramer.

(b) By deriving the counterpart of (48), compare the strength of cooperativity in dimers and tetramers. Are your results intuitively reasonable?

Supplement: Solution of equations (4.15) and (4.16)

In time-dependent problems it is frequently useful first to search for steady (i.e., time-independent) solutions and then to examine the deviation from the steady state. Denoting the steady state of C by the constant $\bar{C}$, we see from (4.15) that $\bar{C}$ satisfies

$$k_1(E_0 + C_0 - \bar{C})S_0 - (k_{-1} + k_2)\bar{C} = 0. \tag{1}$$

Simple algebra yields formula (4.18a) for $\bar{C}$.

The deviation $C'(t)$ from the steady state is defined by

$$C'(t) = C(t) - \bar{C}. \tag{2}$$

Substituting into (4.15)

$$C(t) = C'(t) + \bar{C}, \quad dC/dt = dC'/dt, \qquad\qquad (3a,b)$$

we obtain as the equation for C' [using (1)],

$$dC'/dt = -kC', \quad \text{where } k \equiv k_1 S_0 + k_{-1} + k_2. \qquad\qquad (4a,b)$$

Equation (4a) has the solution [compare equation (A4.2)]

$$C' = C^* \exp(-kt).$$

Here, C^* is a constant that must be determined by the initial conditions. To this end we first use (3a) to find that the original concentration C is given by

$$C(t) = C^* \exp(-kt) + \bar{C}.$$

The initial condition $C(0) = C_0$ thus requires

$$C_0 = C^* + \bar{C}, \quad \text{i.e., } C^* = C_0 - \bar{C},$$

which at once yields the final result (4.17).

5

The chemostat

Construction and analysis of a model for the chemostat is the goal of this chapter. The chemostat is a device that enables us continuously to grow and harvest bacteria. The consequent "continuous culture" of bacteria is in contrast with "batch culture," in which a fixed quantity of nutrient is supplied and bacteria are harvested after a certain growth period.

A schematic view of the chemostat is provided in Figure 5.1. We shall assume that all materials necessary for growth are supplied in abundance, except for one **critical nutrient** that is typically in somewhat short supply. A solution containing the critical nutrient is supplied to the growth chamber, and an equal volume (per unit time) of solution containing bacteria and partly consumed nutrient is removed. The chemostat is stirred to keep conditions uniform.

Virtually all the information necessary for the construction of a mathematical model of chemostat operation has been supplied. The reader is invited to think about the construction of such a model before reading further. A few additional assumptions must be made; the simplest reasonable possibilities should be used, as is the case in all first attempts at model building. We shall shortly sketch the classic theory of the chemostat, due to Novick and Szilard (1950) and Monod (1950). Part of our presentation is similar to that of Rubinow (1975). The reader probably will have little difficulty in following the discussion, but congratulations are in order if he or she has been able to make significant progress toward the model construction. On a number of occasions the author has presented the facts in the previous paragraph to a class and then asked for suggestions on how to proceed. In every case the model was constructed only after about two hours of class debate, and a certain number of hints. This reflects the fact that construction of mathematical models is an art that requires considerable experience. But it is not nearly so difficult to learn to "read" the equations proposed by someone as a mathematical representation of a biological situation, to see what they imply, and hence to judge the suitability of the model. The acquisition of a good

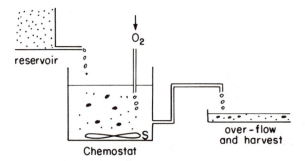

Figure 5.1. Schematic diagram of a chemostat with nutrient supplied from a reservoir, oxygen being bubbled in, and the chemostat chamber being agitated by a stirrer S to keep concentration uniform. Heavy dots represent bacteria. The overflow contains partially used nutrient and the bacterial harvest.

measure of such "reading ability" is one goal of the instruction that we are attempting to impart here.

Choice of variables

We wish to follow the behavior of the chemostat as time passes. In some instances it is appropriate to assume that time progresses in discrete units. For example, yearly time intervals are natural when considering populations of insects that emerge each spring, lay eggs in the fall, and then die in the winter. Here it seems more natural to think of time as an independent variable that continuously increases (but see Exercise 12). The constituents of some systems vary from place to place, necessitating the introduction of further independent variables – the spatial coordinates x, y, and z. Here the stirring creates conditions that can be regarded as spatially homogeneous, so that the time t is the only independent variable required.

Having decided on independent variables, the next step is to select dependent variables, the "unknowns." One of these is $N(t)$, the total number of bacteria in the chemostat at any time t. The other is $C(t)$, the concentration of critical nutrient at time t. Given this selection (which is not obvious to beginners in the art of modeling), we must now try to write some mathematical description that will allow us to compute $N(t)$ and $C(t)$. Here and in many other instances this description takes the form of differential equations that describe the factors that cause the dependent variables to change.

For differential equations to be appropriate, the functions in question must, of course, have derivatives. Here N is an integer, so that its graph

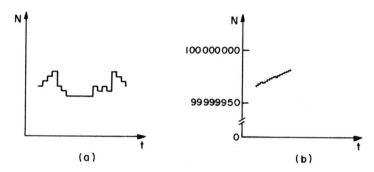

Figure 5.2. Two views of the number of bacteria N as a function of time. (a) A "close up," clearly showing the birth and death of individual bacteria. (b) A more rapidly varying scale N; the steps in the graph are hardly noticeable.

must have the steplike character of Figure 5.2(a), and the derivative is either zero or nonexistent. Nevertheless, because our equations necessarily provide approximate descriptions of the true biological situation, there may be no harm in considering an approximate dependent variable. Indeed, the number of bacteria in the chemostat is very large, and the death or birth of a single bacterium is of no interest to us. Consider the graph of Figure 5.2(b), where a somewhat realistic scale has been provided for N. The jumps in the curve are virtually undetectable. It is thus reasonable to replace the "true" steplike function $N(t)$ by a smooth approximation (to which we shall apply the same letter N).

We now assume that $N(t)$ and $C(t)$ are smooth functions with as many derivatives as we care to calculate. Strictly speaking, as we have pointed out, this is "wrong." But here and in other modeling situations it is important to realize that a model cannot be dismissed (as untrained people are wont to do) merely because it is wrong. Essentially all models of physical situations contain errors, called "simplifications." The question is, Are these simplifications so drastic or foolish that the resulting model is indeed without validity, or can the model provide useful results in a given situation? Here, for example, with inevitable errors in measurement of bacterial and chemical concentrations, the individual bacteria and molecules that inhabit the chemostat are essentially undetectable; so it seems permissible (indeed, wise) to blur their contributions to the state of the system.

Differential equations and initial conditions

There are two major contributions to the change in the total bacteria number N: the net birth of the bacteria (births minus deaths) and the

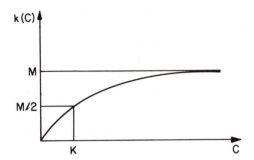

Figure 5.3. The bacterial growth rate k as a function of the critical nutrient concentration C. When $C = K$, k is half of its asymptotic value M.

washing of bacteria out of the chemostat. The rate of the latter step is easy to characterize. Suppose that Q units (e.g., cubic centimeters) of fluid drip out of the chemostat per unit time. The number of bacteria per unit volume is N/V, where V is the volume of the chemostat. Thus, bacteria leave the system at a rate QN/V. Because the rate of change of N is given by the derivative dN/dt, we can write

$$dN/dt = (\text{net births}) - (QN/V). \tag{1}$$

Let k be the average bacterial birthrate. In the absence of other effects, the simplest equation to assume is $dN/dt = kN$, where k is a constant. This yields $N = N_0 \exp(kt)$, so that k can be determined by the doubling time T:

$$2N_0 = N_0 \exp(kT), \quad \text{i.e., } k = (\ln 2)/T.$$

When the bacterial population gets large, the net **birthrate** (or **growth rate**) k will not remain constant. It may be reduced by the accumulation of poisons, for example. Such effects are often modeled by making k a decreasing function of the population size N. Here the population of bacteria is limited because bacteria are washed out of the chemostat, so that the dependence of k on N need not be taken into account in a first modeling attempt. What is important, however, is the dependence of k on the critical nutrient concentration C. When no critical nutrient is present, $k = 0$. Presumably, as critical nutrient concentration rises, the growth rate increases. One foresees a limit to such an increase, for there is a limit to how fast the bacteria can absorb any nutrient. Thus, we expect the dependence of k on C to be given by a curve such as that in Figure 5.3. (Actually, the curve probably decreases at very high values of C, but such an effect is secondary, and we shall ignore it.)

At this point, we have completed our formulation of the equation for N:

$$dN/dt = k(C)N - qN. \tag{2}$$

Here we have introduced the convenient abbreviation

$$q = Q/V. \tag{3}$$

The critical nutrient concentration C increases because of inflow into the chemostat, and C decreases because of outflow and consumption by the bacteria. Denote by C_i the concentration of critical nutrient in the incoming fluid. Because the fluid enters at the constant rate Q, the *amount* of nutrient increases from inflow at the constant rate QC_i. The fluid that leaves at time t is at concentration $C(t)$. At this time, therefore, the efflux of material from the chemostat decreases the net amount of critical nutrient at the rate QC. Finally, let us assume that some function $r(C)$ describes the rate at which an individual bacterium consumes C. Taking account of the fact that there are N bacteria in the chemostat at time t, we are led to the following equation for the change in the total amount of critical nutrient:

$$d(CV)/dt = QC_i - QC - Nr(C). \tag{4}$$

What form should we assume for the consumption function $r(C)$? Obviously, $r = 0$ when $C = 0$. Moreover, r should saturate (approach an asymptote, in mathematical language) when C becomes large, for there must be a limit as to how fast the nutrient can be absorbed. The function $r(C)$ thus has the same general shape as the growth-rate function $k(C)$ depicted in Figure 5.3. In fact, it is reasonable and convenient to assume that the growth rate k is *proportional* to the nutrient uptake rate r. This is almost certainly appropriate at sufficiently low values of C, when both $k(C)$ and $r(C)$ generally are proportional to C (Exercise 5). Later we can determine what changes in our results will occur if this assumption is not made (Exercise 6). We therefore assume that

$$k(C) = yr(C), \tag{5}$$

where y is a constant. We now use (5) to replace r in (4). Remembering that V is a constant, we write $d(CV)/dt = V(dC/dt)$, divide through by V, and recall that $q = Q/V$. This gives as our final version of the equation for the critical nutrient concentration

$$dC/dt = q(C_i - C) - N(yV)^{-1}k(C). \tag{6}$$

Equations (2) and (6) describe how the number of bacteria N and the nutrient concentration C change with time. We should thus be able to calculate the **state** of the chemostat system at any time (i.e., the values of the dependent variables N and C) provided that we know how the system was constituted at some initial time. Let us call this time $t=0$. We assume that the critical nutrient initially was present at concentration C_0 and that a certain number of bacteria N_0 were inoculated into the chemostat as it was started. This gives the initial conditions

$$N(0)=N_0, \quad C(0)=C_0. \tag{7}$$

The differential equations (2) and (6) and the initial conditions (7) constitute a mathematical model for the chemostat. The model contains the parameters q, C_i, y, V, N_0, C_0, and the function $k(C)$; these must be prescribed before any numerical results can be obtained. It will be convenient later to choose a specific expression for the function k. The simplest function that starts at the origin and saturates is given by the "Michaelis–Menten" expression

$$k(C)=MC/(K+C), \quad C\geqslant 0. \tag{8}$$

Here, M and K are constants. Note that [as was pointed out in connection with the analogous equation (4.29)] M is the value of k that is approached as $C\to\infty$, whereas K is the half-saturation constant, in that $k=M/2$ when $C=K$ (see Figure 5.3).

It is worth collecting all the equations (2), (6), (7), and (8) of our explicit mathematical model:

$$\frac{dN(t)}{dt}=\frac{MN(t)C(t)}{K+C(t)}-qN(t), \tag{9a}$$

$$\frac{dC(t)}{dt}=q[C_i-C(t)]-\frac{N(t)MC(t)}{y[K+C(t)]V}, \tag{9b}$$

$$N(0)=N_0, \quad C(0)=C_0. \tag{9c,d}$$

To retain biological meaningfulness, we must add the conditions $N\geqslant 0$, $C\geqslant 0$. Note that the model contains eight parameters q, C_i, y, V, N_0, C_0, M, and K.

At the beginning of this chapter the reader was asked to attempt the unaided construction of a mathematical model of a chemostat. If such an effort was made, the model (9) will not be dismissed as obvious.

We now turn to a mathematical analysis of the model. Here, too, it is recommended that the reader attempt to anticipate the results. We wish to know, for example, if the chemostat will operate as designed. That is, after an initial transient period, will the chemostat settle down to a steady mode of operation in which constant numbers of bacteria per unit time are produced? Is it certain that steady-state production of bacteria will always take place? If so, about how long will it take to achieve a steady state? What parameter values give the maximum production? Is it possible that the bacteria population might oscillate in time? If it does, approximately what is the period of the oscillation? What is the amplitude? Under what conditions will an oscillation occur? If the reader can confidently answer all questions such as these, then there is no need to consider a mathematical model. Otherwise, the use of a mathematical model will doubtless be a useful adjunct to an experimental program.

Steady solutions

Will the chemostat settle down to a steady state of operation? That is, does $N \to$ constant and $C \to$ constant, whatever the initial conditions, as $t \to \infty$? This is a difficult question to answer. But considerable insight can be gained into this problem, and similar problems, if we ask a somewhat different (and much simpler) question: Can the system remain forever in exactly the same **steady state**? That is, do the original equations (2) and (6) admit the following very simple solution?

$$N(t) = \bar{N}, \quad C(t) = \bar{C}, \quad \bar{N} \text{ and } \bar{C} \text{ constants.} \tag{10}$$

Because the derivative of a constant is zero, (10) will hold if and only if

$$k(\bar{C})\bar{N} - q\bar{N} = 0, \quad q(C_i - \bar{C}) - \bar{N}(yV)^{-1}k(\bar{C}) = 0. \tag{11a,b}$$

The algebraic equation (11a) has two solutions: one if $\bar{N} = 0$, and another if $k(\bar{C}) = q$. Let us consider these possibilities one at a time.

If $\bar{N} = 0$, then (11a) is satisfied, whereas (11b) is satisfied if and only if $\bar{C} = C_i$. This solution is so "trivial" that it is easily overlooked, but it makes perfect biological sense. The chemostat can run forever with no bacteria inside, and hence with no change in the concentration of critical nutrient.

We must now ask about the solutions to the equation $k(\bar{C}) = q$. Does this equation always have a solution? Could it have more than one solution? The answer (as is so often the case) is easily seen by drawing some graphs. We see from Figure 5.4 that $k(\bar{C}) = q$ has *no* solutions if $q \geqslant M$ and exactly one solution if $q < M$. Again, this makes sense biologically. As far as the bacteria in the chemostat are concerned, "washout" at rate qN is the same as death, for the washed-out bacteria disappear. If the

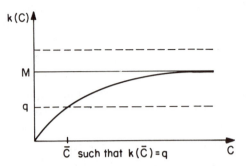

Figure 5.4. Demonstration of the fact that the equation $k(C)=q$ has a unique solution when $q<M$ and no solution when $q\geqslant M$.

washout rate q is greater than the maximum birthrate M, then there can be no nonzero steady population level of bacteria in the chemostat.

Once $\bar{C}$ is obtained, by solving $k(\bar{C})=q$, then a unique value of $\bar{N}$ can be obtained by solving (11b). For this value to be positive, q must be such that $\bar{C}<C_i$.

To summarize, there are two possible steady-state solutions of the equations (2) and (6). One solution is

$$\bar{N}=0, \quad \bar{C}=C_i. \tag{12a}$$

Corresponding to every positive flow parameter q that is less than the maximum value of $k(C)$ and is also small enough so that $C_i>\bar{C}$, there is a second solution $\bar{C},\bar{N}$, where

$$k(\bar{C})=q, \quad \bar{N}=Vy(C_i-\bar{C}). \tag{12b}$$

If we use (8), we can find explicit formulas for the solution (12b):

$$\text{If } q<M, \quad \bar{C}=\frac{qK}{M-q}, \quad \bar{N}=Vy\left(C_i-\frac{qK}{M-q}\right). \tag{13a,b}$$

Here we see algebraically that $\bar{C}>0$ requires

$$q<M,$$

whereas $\bar{N}>0$ if

$$C_i>\frac{qK}{M-q}.$$

To transform the latter inequality into a condition on the flow-rate parameter q, we multiply both sides by $M-q$. Because this quantity must be positive, as we have just seen, we obtain

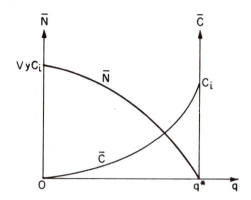

Figure 5.5. Graphs of the steady-state bacteria number ($\bar{N}$) and critical nutrient concentration ($\bar{C}$) as functions of the flow-rate parameter q [see equations (13) and (14)].

$$C_i M - C_i q > qK,$$

or

$$q < q^*, \quad \text{where } q^* = \frac{M}{1 + (K/C_i)}. \tag{14}$$

Note that $q^* < M$, so that if $\bar{N} > 0$, then certainly $\bar{C} > 0$.

Figure 5.5 depicts graphs of the steady-state nutrient concentration $\bar{C}$ and bacteria number $\bar{N}$ as functions of the flow parameter q. It is seen that as q increases toward q^*, $\bar{C}$ increases to its maximum possible value C_i, while $\bar{N}$ decreases toward zero.

If the chemostat is regarded as a factory for producing bacteria, then it is natural to inquire as to its optimum output. One possible quantity to maximize is the bacterial production rate $q\bar{N}$. This is zero when $q = 0$, and also when $q = q^*$ (and $\bar{N} = 0$). There must be an optimum intermediate flow rate.

Problem: Find the flow rate q that maximizes $q\bar{N}$.
Solution: From (13b),

$$q\bar{N} = Vy\left(qC_i - \frac{q^2 K}{M - q}\right).$$

The derivative $\partial(q\bar{N})/\partial q$ vanishes when

$$C_i - \frac{(M - q)(2qK) + q^2 K}{(M - q)^2} = 0. \tag{15}$$

Simplifying, we observe that (15) holds if

$$q^2(C_i + K) - 2Mq(C_i + K) + M^2 C_i = 0. \tag{16}$$

It is an important observation that (16) can be further simplified if we divide through by $C_i + K$ and use (14), yielding

$$q^2 - 2Mq + Mq^* = 0. \tag{17}$$

On solving (17), we find that

$$q = M - [M(M - q^*)]^{1/2}. \tag{18}$$

Because $q^* < M$, the quantity whose root must be extracted is positive. We have not written a plus sign before the square-root term, because the plus sign corresponds to the value of q that is larger than M and is therefore of no biological interest. But we must verify that the other root is not only less than M but also less than q^*.

Is $M - [M(M - q^*)]^{1/2} < q^*$?   I.e., is $M - q^* < [M(M - q^*)]^{1/2}$?

Is $(M - q^*)^2 < M(M - q^*)$?   Is $M - q^* < M$?   Yes. (19)

The maximum flow rate is not the only possible optimization that may be sought. For example, we might wish to maximize profit, given certain prices to be paid for nutrient and to be received for bacteria (Exercise 7).

Dimensionless variables

At some stage in the examination of a nontrivial mathematical model it is often wise to introduce dimensionless variables, if only to simplify the notation by the consolidation of some parameters (see Appendix 7). We must divide the existing variables by parameters of the same dimensions. A way in which this can be done in the present case is to measure time in multiples of $1/q = V/Q$, the time it takes to fill (or empty) the chemostat at rate Q. The first term of (9a) will simplify if we define a dimensionless chemical concentration c by $C(t) = Kc(t)$, that is, if we measure concentration in units of the half-saturation constant K. And it is natural to measure N in terms of the initial inoculant size N_0. With the dimensionless variables

$$\tau = tq, \quad c = C/K, \quad n = N/N_0, \tag{20a,b,c}$$

the system (9) becomes [Exercise 1(a)]

◆◆
$$\frac{dn}{d\tau} = \frac{\mu nc}{1 + c} - n, \quad \frac{dc}{d\tau} = \xi - c - \nu \frac{\mu nc}{1 + c}, \tag{21a,b}$$

$$n(0) = 1, \quad c(0) = c_0. \tag{21c,d}$$

Instead of the eight parameters of (9), the dimensionless equations (21) contain the four dimensionless parameters

◆◆ $$\mu \equiv M/q, \quad \xi = C_i/K, \quad \nu = N_0/yKV, \quad c_0 = C_0/K. \qquad \text{(22a–d)}$$

The dimensionless steady solutions satisfy

$$\bar{n} = 0, \quad \bar{c} = \xi, \qquad \qquad \text{(23a,b)}$$

or

$$\frac{\mu \bar{c}}{1 + \bar{c}} = 1, \quad \xi - \bar{c} - \frac{\nu \mu \bar{n} \bar{c}}{1 + \bar{c}} = 0, \qquad \text{(24a,b)}$$

so that [Exercise 1(b)]

$$\bar{n} = \frac{\xi}{\nu} - \frac{1}{\nu(\mu - 1)}, \quad \bar{c} = \frac{1}{\mu - 1}. \qquad \text{(25a,b)}$$

The positivity of the solutions in (25) is assured by the following dimensionless version of (14):

$$\mu > 1 + \xi^{-1}. \qquad \text{(26)}$$

Stability of the steady states

Under certain conditions we have two possible steady states: (23) and (25). To which of these, if either, will the system tend? This is not an easy question to answer in general, but we can answer it rather simply for solutions that start *near* one of the steady states. The required "stability analysis" is always worthwhile as part of an attempt to understand the behavior of a system of ordinary differential equations (see Appendix 5). Our calculations will be facilitated by the relatively uncluttered form of the dimensionless equations (21), but except for slightly increased algebraic complication, they could just as well have been carried out on the original equations (9) (Exercise 3).

The given equations (21) have the form

◆◆ $$dn/d\tau = f(n, c), \quad dc/d\tau = g(n, c), \qquad \text{(27a,b)}$$

where

$$f(n, c) = \frac{\mu n c}{1 + c} - n, \quad g(n, c) = \xi - c - \frac{\nu \mu n c}{1 + c}.$$

As is discussed in Appendix 5, the procedure is to introduce deviations $n'(t)$ and $c'(t)$ from a given steady state $(\bar{n}, \bar{c})$ by

◆◆ $$n' = n - \bar{n}, \quad c' = c - \bar{c}. \qquad \text{(28a,b)}$$

Assuming that the quantities n' and c' are small, we can utilize the Taylor approximation (A2.8) to obtain the approximate (linear) equations

◆◆
$$dn'/d\tau = An' + Bc', \quad dc'/d\tau = Cn' + Dc', \tag{29a,b}$$

where

$$A \equiv \left.\frac{\partial f(n,c)}{\partial n}\right|_{\substack{n=\bar{n}\\c=\bar{c}}} = \frac{\mu\bar{c}}{1+\bar{c}} - 1, \quad B = \left.\frac{\partial f(n,c)}{\partial c}\right|_{\substack{n=\bar{n}\\c=\bar{c}}} = \frac{\mu\bar{n}}{(1+\bar{c})^2}, \tag{30a,b}$$

$$C = \left.\frac{\partial g(n,c)}{\partial n}\right|_{\substack{n=\bar{n}\\c=\bar{c}}} = -\frac{\nu\mu\bar{c}}{1+\bar{c}}, \quad D = \left.\frac{\partial g(n,c)}{\partial c}\right|_{\substack{n=\bar{n}\\c=\bar{c}}} = -1 - \frac{\nu\mu\bar{n}}{(1+\bar{c})^2}. \tag{30c,d}$$

Let us first consider the steady-state point (25). Using (24a) for preliminary simplification of the expressions in (30) (not an essential step, but one that makes calculations easier), we find (Exercise 4) that

$$A = 0, \quad B = \theta, \quad C = -\nu, \quad D = -1 - \nu\theta. \tag{31}$$

Here,

$$\theta = \frac{(\mu-1)^2}{\mu\nu}\left(\xi - \frac{1}{\mu-1}\right) \tag{32}$$

is a positive constant, by (26). From (A5.16), the next step is to calculate β and γ:

$$\beta \equiv -(A+D) = 1 + \nu\theta, \quad \gamma = AD - BC = \nu\theta. \tag{33}$$

Because

$$\beta > 0, \quad \gamma > 0, \quad \beta^2 - 4\gamma = (\nu\theta+1)^2 - 4\nu\theta = (\nu\theta-1)^2 > 0,$$

we see from Figure A5.9 that the steady-state point (25) is a stable node. Suppose that circumstances are such that the steady state in question exists [i.e., (26) holds]. Then our stability result implies that if the chemostat starts near this steady state, conditions in the chemostat will automatically adjust themselves so that steady-state conditions (with a nonzero bacterial population) will be approached even more closely. (This is a homeostatic property that is typical of biological systems. The investigation of homeostasis mandates stability analyses in many biological problems.)

Having examined (25), we turn our attention to the other steady state (23), wherein there are no bacteria in the chemostat. Here (Exercise 4),

$$A = \frac{\mu\xi}{1+\xi} - 1, \quad B = 0, \quad C = -\frac{\nu\mu\xi}{1+\xi}, \quad D = -1.$$

$$\beta = 2 - \frac{\mu\xi}{1+\xi}, \quad \gamma = 1 - \frac{\mu\xi}{1+\xi}. \tag{34}$$

Perusal of Figure A5.9 leads us to conclude that there are two possibilities. If

$$1 < \frac{\mu\xi}{1+\xi}, \quad \text{so that } \gamma < 0, \tag{35a,b}$$

then (23) is a saddle point. Alternatively, if (35a) is reversed, then

$$\gamma > 0, \quad \beta > 0, \quad \beta^2 - 4\gamma = (\gamma+1)^2 - 4\gamma = (\gamma-1)^2 > 0,$$

and (23) is a stable node.

We must now relate the conditions just obtained to the other inequalities that we have derived. In doing so, we observe that (35a) is in fact identical with (26). Consequently, either there is a single stable steady state (with no bacteria) or there are two steady states – one with no bacteria, and one in which the chemostat is operating properly. In the latter case, the steady state (25) with bacteria is stable, and the no-bacteria state (23) is unstable.

Qualitative behavior and the phase plane

Determination of steady states and their stability is a relatively straightforward task that it is almost always worth carrying out for any dynamic model (i.e., for any set of time-varying equations). Accomplishment of this task is a major step toward the objective of ascertaining the model's qualitative behavior for all possible sets of parameters and initial conditions.

For the present problem, for every (biologically meaningful) parameter value there is exactly one stable steady state. Thus, it is natural to conjecture that the chemostat will approach this state from all initial conditions. This would mean, from (14), that if the flow rate q were larger than the critical value q^*, then in whatever manner the chemostat was started, all the bacteria would eventually disappear. But if $q < q^*$, there would be a strong homeostatic tendency. No matter how far the chemostat was started from its steady-state operating conditions, the "proper" steady state would be approached as time passed.

The conjecture we have just made about the overall or **global behavior** of the chemostat is quite reasonable, and it is possible to terminate our theoretical investigation at this point. But additional effort will make the conjecture firm, in the face of such questions as Could there be an oscillatory solution as well as a steady state? Further light will be shed on the operation of the system, and we shall obtain valuable practice in thinking about the phase plane.

Let us follow the steps suggested in Appendix 5 to conjecture the behavior of the various trajectories in the phase plane. We shall plot the bacterial population n on the ordinate (vertical axis) and the critical substrate concentration c on the abscissa (horizontal axis), in deference to the tendency to view the bacterial population as a function of its nutrient supply. (In fact, it is wise to keep firmly in mind that n and c are mutually interdependent.)

Horizontal tangents in the governing system (21a,b) occur when

$$\frac{dn}{d\tau} = 0, \quad \text{i.e., when } n\left(\frac{\mu c}{1+c} - 1\right) = 0.$$

Thus, **horizontal nullclines** are the straight lines $n = 0$ and $c = (\mu - 1)^{-1}$. The **vertical nullcline** (where $dc/dt = 0$) is the curve

$$\xi - c - \frac{\nu \mu n c}{1+c} = 0, \quad \text{i.e., } n = \frac{(\xi - c)(1 + c)}{\nu \mu c}.$$

This curve approaches $+\infty$ as c decreases to zero, and it cuts the c axis when $c = \xi$. In checking for intersections of vertical and horizontal nullclines (the steady-state points), we observe that there are two cases, depending on the relation between ξ and $(\mu - 1)^{-1}$. This is shown in Figure 5.6, where we have also indicated the directions of the trajectories on the nullclines, as well as their directions on the vertical axis (Exercise 8).

Knowing the information we do, it is now possible to sketch the trajectories of Figure 5.7 with quite good confidence, for there seem to be no other reasonable alternatives.

Note how the typical trajectory labeled A moves upward and to the left in the beginning, because the initially large nutrient supply permits rapid growth. This, in turn, leads to ever more rapid depletion of bacterial nutrient. Eventually a time is reached at which the population begins to decline; at that time, bacterial density is maximal, at a point of horizontal tangency.

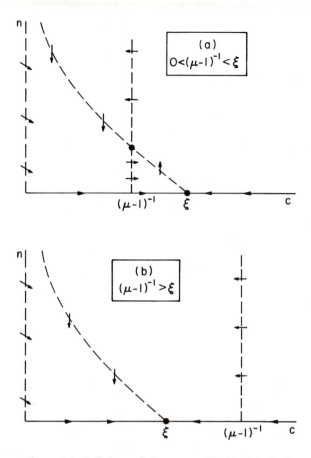

Figure 5.6. Nullclines of the system (21a,b) with the intersections of vertical and horizontal nullclines (steady-state points) marked by heavy dots. Arrows indicate the directions of trajectories on the nullclines and on the axes.

It is a recurrent theme of this book that dynamic systems should be analyzed by examining steady-state points and their stability, conjecturing qualitative behavior, and testing the conjectures by means of computer simulations. For the difference equations of Chapters 2 and 3, such simulations were relatively easy to carry out. They are more difficult for differential equations, but preprogrammed "packages" are becoming available that make phase-plane analysis easier and easier. One such package has been written by Professor G. Odell, of the Department of Mathematical Sciences, Rensselaer Polytechnic Institute, and

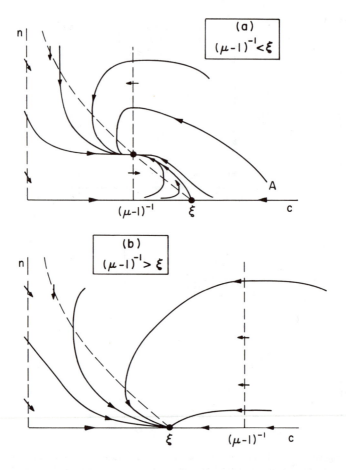

Figure 5.7. Trajectories consistent with the information of Figure 5.6. The trajectory marked *A* is discussed in the text.

was kindly used by him to produce Figure 5.8. This figure provides gratifying verification of the qualitative behavior that was sketched in Figure 5.7.[1]

The author is not expert with the computer. In an hour of need, he has been able to induce Professor Odell to produce some desired computer simulations. Similarly, when in need of assistance, the reader can doubtless enlist the services of a sympathetic computer virtuoso.

1 Results of an honest test are being reported. Figure 5.7 was sketched *before* Figure 5.8 was constructed.

Final remarks

We have answered the various questions we posed concerning the qualitative behavior of the chemostat. Our calculations give strong evidence that the chemostat always approaches a steady state. The approach is nonoscillatory, because steady-state points are never foci.

Although eight parameters appear in our model (9), we have seen that the major qualitative features of the system depend on whether or not the two dimensionless parameters μ and ξ obey inequality (26). Figure 5.9 sums up the situation. In this figure we have chosen to present the results in terms of μ^{-1}, not μ, because the former is proportional to the most easily adjustable parameter, the flow rate Q.

In terms of the original parameters, the condition of Figure 5.9 for the final state to contain no bacteria becomes the following constraint on the flow parameter $q \equiv Q/V$ [see equation (14)]:

$$q > MC_i/(K+C_i). \tag{36}$$

To interpret this condition, consider the (dimensional) bacteria equation (9a):

$$dN/dt = NMC(t)/[K+C(t)] - qN.$$

In this equation, the flow parameter q can be regarded as playing the role of a death rate, and $MC(t)/[K+C(t)]$ is a nutrient-dependent birthrate. The largest birthrate that can occur in the chemostat is when the nutrient concentration C takes on its maximum possible value of C_i. If and only if this maximum birthrate exceeds the "death rate" q will a nonzero population eventually emerge.

None of our major conclusions may seem surprising in retrospect, but few are generally able to predict our findings without mathematical analysis. Indeed, the chemostat field began with the fundamental theoretical papers of Novick and Szilard (1950) and Monod (1950). The theory has become more necessary as biologists use the chemostat to study more complicated matters, such as competition between different organisms (Slobodkin, 1961) (see Exercise 10) and the effects of temporally varying conditions (Frisch and Gotham, 1977; Smith, 1981).

Experiments have indicated that the basic theory can indeed capture the main features of the experimental phenomena (Rubinow, 1975, Section 1.7), but, as outlined in Exercise 13, recent measurements have provided surprises that require extensions of the analysis.

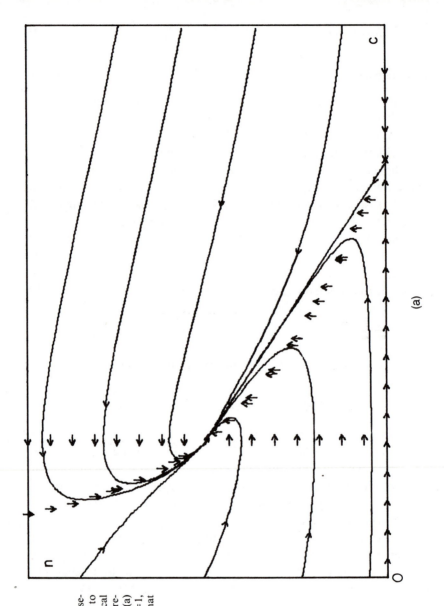

Figure 5.8. Computer-drawn phase-plane diagrams corresponding to Figure 5.7. Horizontal and vertical nullclines are indicated by corresponding loci of short arrows. (a) Solutions of (21) with $\xi = 3$, $\nu = 1$, $\mu = 2$. (b) Same as (a) except that $\mu = 1.25$.

(a)

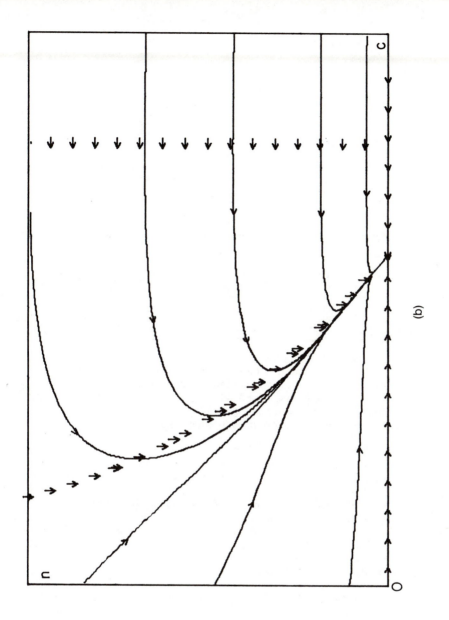

(b)

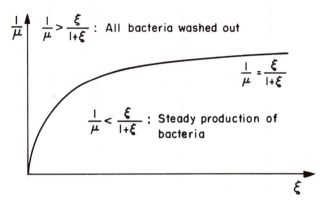

Figure 5.9. Parameter plane indicating domains of different qualitative behaviors of solutions to (21). As in (22), $\mu^{-1} \equiv q/M$ and $\xi \equiv C_i/K$.

Exercises

1. (a) Verify equation (21).
   (b) Verify equations (23)–(26) both by proceeding directly from (21) and by translating earlier formulas into dimensionless variables.
   (c) Verify that the graphs of $\bar{N}$ and $\bar{C}$ are respectively concave down and concave up, as depicted in Figure 5.5.
2. Instead of equation (20), introduce the variables

   $$\tau = tM, \quad c = C/C_i, \quad n = N/N_0.$$

   (Note that these variables are dimensionless.) Find the equations corresponding to (21) in terms of the dimensionless parameters $\alpha \equiv q/M$, $\beta \equiv K/C_i$, $\gamma = N_0/yVC_i$, and $\delta = C_0/C_i$. Express α, β, γ, and δ in terms of the dimensionless parameters of (22), and thereby show that the latter could still be used to characterize the problem.
3. For practice, repeat the stability calculations of the text using the original equations (9) rather than the dimensionless version (21). Show that the results of the two approaches are the same.
4. Verify equations (31) and (34).
† 5. Show that assumption (5) is "usually" appropriate, at least for sufficiently small values of C. Describe circumstances, using a specific example, under which (5) would not be a valid approximation, even for small C.
6. Consider the case in which assumption (5), that k and r are proportional, is not made.
   (a) Show that there is essentially no change in the steady states.
   (b) Discuss the stability of the steady states.
   (c) Describe the effect of this generalization, if any, on the qualitative behavior.
7. How should the chemostat be run to maximize profit if a certain sum is received per bacterium, taking into account the cost of critical nutrient? Consider two cases:

(a) In the first case, no nutrient that enters the chemostat is recoverable.
(b) In the second case, all nutrient that leaves the chemostat can be recycled.

8. Carry out the remaining analysis necessary to obtain Figure 5.6.
9. (a) Find the general solution of the linear system given by (29) and (31).
   (b) Keeping in mind that $\exp(-qt)$ decays appreciably in a time of order q^{-1} [see equation (4.20)], show that if $(1+\theta)^2 < 4\theta\nu$, then $(1+\theta)/2q$ estimates the time for the chemostat to approach nontrivial steady conditions, provided that the system is not started too far from such conditions.
   (c) Show that if

   $$(1+\theta)^2 > 4\theta\nu,$$

   then the corresponding time is $\max(m_1 q^{-1}, m_2 q^{-1})$, where $-m_1$ and $-m_2$ are the roots of $m^2 + (1+\theta)m + \theta\nu = 0$.

10. This problem concerns the competition between two species in a chemostat (or, equivalently, two species subject to indiscriminate predation or harvesting). Let $N_1(t)$ and $N_2(t)$ be the number of individuals in the chemostat.
    (a) We shall employ the equations

    $$\frac{dN_1}{dt} = r_1 N_1 \left(1 - \frac{N_1 + \alpha_{12} N_2}{K_1}\right) - qN_1,$$

    $$\frac{dN_2}{dt} = r_2 N_2 \left(1 - \frac{N_2 + \alpha_{21} N_1}{K_2}\right) - qN_2.$$

    Discuss the assumptions that are implicit in these equations.
    (b) Show that coexistence of the two species is possible (Slobodkin, 1961) if

    $$K_1\left(1 - \frac{q}{r_1}\right) > \alpha_{12} K_2\left(1 - \frac{q}{r_2}\right), \quad K_2\left(1 - \frac{q}{r_2}\right) > \alpha_{21} K_1\left(1 - \frac{q}{r_1}\right),$$

    where $q < r_1$, $q < r_2$.
    (c) Show that if $q = 0$, then species 1 will always win in competition if $K_1 > \alpha_{12} K_2$, $K_2 > \alpha_{21} K_1$.
    (d) Given that $\alpha_{12}\alpha_{21} < 1$, $r_1 < r_2$, show that, in spite of the intrinsic superiority of species 1, if

    $$\frac{K_1 - \alpha_{12} K_{12}}{K_1 r_1^{-1} - \alpha_{12} K_2 r_2^{-1}} < q < \frac{\alpha_{21} K_1 - K_2}{\alpha_{21} K_1 r_1^{-1} - K_2 r_2^{-1}},$$

    then coexistence is possible. (Ecologically speaking, this shows that indiscriminate predation can enhance the possibilities of species coexistence.)
    (e) Find conditions permitting coexistence when $q = 0$ but not when $q > 0$.

11. An extremely simple model for the number of bacteria $N(t)$ in some natural environment is

    $$\frac{dN}{dt} = \frac{MCN}{C+k} - DN, \quad \frac{dC}{dt} = -\frac{MCN}{Y(C+k)} + I,$$

    where $M, k, d, y,$ and I are positive constants.

†   (a) Explain what has been assumed in this model. In particular, what are the meanings of all the various constants?

(b) Show that there is at most one possible steady state.

(c) Examine the stability of the steady state.

(d) Conjecture qualitative behavior when a single steady state exists, and sketch the corresponding phase plane.

(e) Conjecture qualitative behavior when no steady state exists, and sketch the corresponding phase plane.

† 12. Construct from first principles an alternative to the model of (2) and (6) based on discrete time intervals. (Imagine, for example, that measurements of N and C are made every few minutes.)

13. In our model of the chemostat we assumed that the eating rate r depended on the critical nutrient concentration $C(t)$. But in batch-culture experiments on the unicellular alga *Chlamydomonas reinhardii* it was shown that division continues for many hours after the critical nutrient (nitrate) disappears. A modification of the standard model has therefore been suggested in which r depends on $U(t)$, the average amount of stored nutrient per cell, at time t.

(a) Justify the assumption of

$$\frac{d}{dt}(NU) = rN - qUN,$$

where $N(t)$, r, and q have the same meanings as in the text.

(b) Use equation (2) to deduce the following equation for U:

$$dU/dt = r - kU.$$

For a study of the three equations for $U(t)$, $N(t)$, and $C(t)$ and comparison with experiment, see Nisbet and Gurney (1982).

6

A model of the cellular slime mold cAMP signaling system

This chapter is concerned with the chemical signaling system of the cellular slime mold *Dictyostelium discoideum*. We begin with a brief sketch of the salient biological background.

The slime mold life cycle

Cellular slime molds are found in the soil throughout the world. We can commence consideration of their life cycle with the spore stage. Under favorable conditions the spore will germinate, and an amoeba will emerge from its protective casing. The amoebae move about by virtue of their pseudopods, consuming bacteria for food. They reproduce by fission, with about one division every three hours under ideal conditions. Reproduction continues as long as bacterial food is available. When the food supply is exhausted, the amoebae appear to move about randomly for about eight hours, after which they commence a remarkable aggregation movement toward more-or-less equally spaced collection points.

The nature of aggregation itself and the events that follow it differ somewhat from species to species. We shall concentrate here on *D. discoideum*, the most studied of these organisms. After aggregation, *Dictyostelium* amoebae form a multicellular slug that wanders toward light and more humid conditions. The slug eventually stops, and for several hours goes through a period of "purposeful" internal motion and change. At the end of this time, the slug has transformed itself into a delicate stalk composed of dead, rigid, vacuolated cells, on top of which rests a ball of spores (see frontispiece.). If a spore is transported to a place that is favorable for germination, the whole cycle will begin again [Surveys of slime mold biology can be found in the books of Bonner (1967) and Loomis (1982).]

For decades *D. discoideum* has been the object of intense and intensifying investigation by biologists. Two principal reasons for this interest are already apparent. First, morphogenetic (form-producing) movement is manifested in a striking yet relatively simple way. Moreover, cellular slime molds beautifully exhibit a second major process in developmental biology in which an originally quite homogeneous collection of cells differentiates in a controlled fashion into two cell types: stalk and spore.

The role of cAMP

It was conjectured some time ago that aggregation takes place because the cells are attracted by relatively high concentrations of a chemical that they themselves secrete. Such chemically directed motion is called **chemotaxis**. More recently, the attractant for *D. discoideum* has been identified as cyclic adenosine 3',5'-monophosphate (called cyclic AMP or simply cAMP). It has also been shown that the cells secrete the enzyme **phosphodiesterase** (PDE), which catalyzes the breaking of a bond in cAMP, turning it into 5'-AMP, which is not a chemoattractant for the amoebae.

The discovery that cAMP plays a major role in *D. discoideum* gave great impetus to the study of this organism, for cAMP also has a major function in mammalian physiology as the "universal second messenger" of hormone action. Indeed, E. W. Sutherland received a Nobel Prize in 1971 for showing that after any one of many hormones binds to a corresponding specific site on a target cell, the second step is almost always the same: activation of the enzyme **adenylate cyclase**, which catalyzes the synthesis of cAMP. Because cAMP also plays a vital role in the biology of *D. discoideum*, hopes have been reinforced that an understanding of developmental processes in slime mold will have a considerable degree of general significance. Moreover, the slime mold biologists have been able to take advantage of sensitive assays developed for cAMP that can give quantitative estimates of its concentration, even though this concentration is typically in the range of 10^{-9}–10^{-12} M.

In a typical aggregation pattern in *D. discoideum*, individual amoebae move steadily toward the aggregation center for about a minute and a half, and then make no further progress for a period of approximately three to eight minutes. The repetition of a number of such "step-pause" actions brings the cell to the center. Observations of the whole field of cells reveal that the steps occur in sequence: First the cells nearest the center step inward, then an adjacent ring of cells farther out, then the next outward ring, and so forth. Thus, aggregation takes place in pulses, somewhat spasmodic movements by the individual cells, with outward-moving waves of inward steps. See frontispiece (top): The moving bands of amoebae appear relatively bright.

Shaffer (1962) conjectured that the steps were the result of a pulsatile secretion of attractant that was relayed from one circle of cells to another after a certain delay. This conjecture has been verified by a technique developed by Gerisch and Hess (1974), wherein cells are suspended and continually agitated in liquid culture. Because conditions are uniform, cell development is synchronized. Virtually no cAMP appears in the medium until about eight hours after starvation. Then the assemblage of cells spontaneously begins a periodic secretion of cAMP, with an interval of 5–10 minutes between pulses. But if cAMP is artificially injected into the cell suspension between six and eight hours following starvation, the cells respond after a delay of about half a minute with a much larger cAMP secretion than that to which they have been exposed. It thus appears that pulsatile aggregation can take place because a few cells spontaneously begin a periodic cAMP secretion; the cAMP attracts nearby cells and also stimulates them to *relay* the signal by activating the enzyme adenylate cyclase that catalyzes cAMP synthesis.

There is evidence that cAMP plays other roles, in addition to its function in the relay-chemoattractant mode of aggregation. Pulses of cAMP can induce the synthesis of certain specific proteins related to aggregation, and these pulses may control wavelike chemotactic movements in the slug. More important for present purposes, it appears that cells at the center of the aggregate secrete cAMP at a high steady level. It has been shown that anterior prestalk cells show relatively high cAMP concentrations. It may even be that the interplay of the chemoattractant cAMP and its inhibition by ammonia play a decisive role in directing the slug-fruiting-body transition and the morphogenetic events that lead to the mature fruiting body with its stalk and spores (for references, see Goldbeter and Segel, 1980).

In this chapter we shall fix our attention on one major aspect of the role of cAMP: the fact that there are at least three and probably four different states of cAMP activity around the time of aggregation. These are the following, to summarize our earliest remarks:

(a) Minimal cAMP production. Shortly after starvation, cells secrete little or no cAMP.

(b) Excitable cAMP production (relay). After a few hours, cells can be stimulated by a cAMP signal to secrete cAMP in an amount much larger than that of the stimulus.

(c) Oscillatory cAMP production. After about eight hours, some cells, at least, will secrete a cAMP pulse every few minutes.

(d) Strong steady cAMP production. Some cells at the aggregation center appear to secrete cAMP continuously and relatively strongly.

We shall demonstrate that all these activities, and the transitions between them, can be explained by means of interactions between just a few known features of the cAMP signaling system. Our presentation is based on papers by Goldbeter and Segel (1977, 1980) and Goldbeter, Erneux, and Segel (1978). The mathematical model takes the form of three nonlinear ordinary differential equations, whose analysis will be seen to involve a variety of techniques that we have already employed in simpler problems – determination of steady states and their stability, phase plane, differing qualitative behaviors for different regions of parameter space, and so forth. We shall see, for example, that the oscillatory cAMP production of state (c) is associated with parameter values for which the only steady state of these equations is unstable. By this type of analysis we shall demonstrate that the four behavioral types (a)–(d) are associated with four different domains in a certain parameter plane.

The model

Generally agreed on are the following features of the cAMP signaling system (Loomis, 1982):
   (a) receptors for cAMP on the cell exterior;
   (b) a coupling of the receptors to the enzyme adenylate cyclase that catalyzes cAMP production from the substrate **adenosine triphosphate** (ATP);
   (c) an increase in cyclase activity, and hence an increase in intracellular cAMP concentration, in response to an elevation in the extracellular cAMP concentration;
   (d) secretion of cAMP following an increase in its intracellular concentration;
   (e) hydrolysis (destruction) of cAMP by the enzyme phosphodiesterase.

It is the coupling of step (b) that is the least understood part of the process. We shall assume here perhaps the simplest possible coupling: that in which the receptor on the cell exterior is a regulatory site for the cyclase molecule (whose catalytic site is inside the cell) (Figure 6.1). We assume further that the enzyme is a dimer in a quasi-steady state, of the type for which we have already derived the relevant relation between regulatory concentration and reaction velocity in Chapter 4.

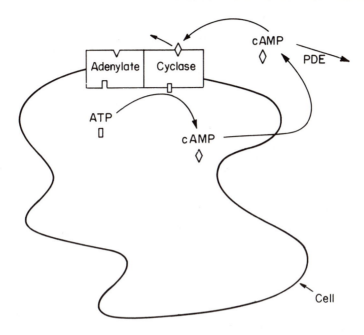

Figure 6.1. Schematic diagram of slime mold cell with enzyme adenylate cyclase catalyzing the transformation of ATP into cAMP. Intracellular cAMP is secreted. In the extracellular medium, cAMP may be hydrolyzed (destroyed as an attractant) by the enzyme phosphodiesterase (PDE), and it also may bind to a site on the cyclase and thereby increase the rate of catalysis.

It could well be that the cyclase has more than two subunits (dimer), say four (tetramer) or six (hexamer). Moreover, it could be that the receptor is a molecule entirely separate from the cyclase, with communication between them mediated by a multistep reaction involving several other chemicals. There are many possible mechanisms for the modification of enzyme activity, from among which, in the absence of further information, we select the simplest. However, our arguments will be unconvincing unless we can demonstrate that the major qualitative features shown by our model remain basically unaltered if different and more complex mechanisms are responsible for the receptor-cyclase coupling. Such a demonstration will be provided later, in the section entitled "Robustness of the results."

We shall now construct what seems to be the simplest mathematical model that includes indispensable features of the cAMP signaling system. To this end we define three dependent variables as functions of the time t:

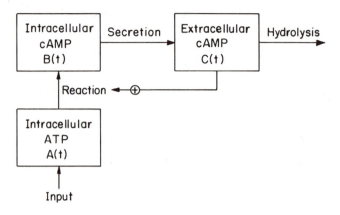

Figure 6.2. Block diagram of the ingredients of the mathematical model, showing positive feedback of extracellular cAMP on the production of intracellular cAMP.

$A(t)$ – the intracellular ATP concentration (the substrate of the reaction),

$B(t)$ – the intracellular cAMP concentration (the product of the reaction),

$C(t)$ – the extracellular cAMP concentration.

The relationships among these three variables are summarized by the **block diagram** of Figure 6.2. Noteworthy is the **positive feedback** (symbolized by a plus sign) whereby extracellular cAMP enhances the synthesis rate of intracellular cAMP. The following are the particular equations that will be assumed:

$$dA/dt = V - W\Phi(A, C), \tag{1a}$$

$$dB/dt = W\Phi(A, C) - k_i^* B, \tag{1b}$$

$$dC/dt = k_i^* h^{-1} B - k^* C, \tag{1c}$$

where

$$\Phi(A, C) = \frac{A}{K_s}\left(1 + \frac{A}{K_s}\right)\left(1 + \frac{C}{K_p}\right)^2 \Big/ \left[L + \left(1 + \frac{A}{K_s}\right)^2\left(1 + \frac{C}{K_p}\right)^2\right]. \tag{1d}$$

Equation (1a) states that the substrate ATP enters the domain of interest at constant "velocity," V, and that the substrate is transformed into cAMP at the rate appropriate to a Monod–Wyman–Changeux product activated dimer with exclusive binding to the R state [see equation (4.60)]. Here, W is the maximum velocity of cAMP production, and, in (1d), K_s is the Michaelis constant of the enzyme for the substrate ATP,

K_p is the dissociation constant for binding of extracellular cAMP with the regulatory site, and L is the allosteric constant.

Equation (1b) provides a term for the appearance of cAMP, a term that, of course, is identical with the term in (1a) for the disappearance of substrate. In addition, it is assumed that secretion of intracellular cAMP is proportional to its concentration, with proportionality constant k_t^*. Perhaps it would be more natural to assume that secretion is proportional to the concentration difference $(B-C)$. Indeed, this assumption was made at the beginning of the investigation reported here. It was found, however, that the simplification $k_t^*(B-C) \approx k_t^* B$ caused little change in the results. The reason is that when secretion is important, it turns out that $C \ll B$.

The last term on the right side of (1c) describes in the simplest possible way the destruction of cAMP by the enzyme phosphodiesterase, at a rate proportional (through the constant k^*) to the cAMP concentration.[1] The other term yields the elevation in extracellular cAMP concentration that is a consequence of cAMP secretion. Here, h is the ratio of extracellular fluid volume to cell volume in the cell-suspension experiments, with whose results we shall compare the theoretical predictions. Inclusion of such a **dilution factor** is clearly necessary when secretion takes place between compartments of differing capacities.

Dimensionless variables

To simplify the equations, we shall introduce dimensionless variables. Equation (1d) makes it natural to refer the concentrations A and C to K_s and K_p, respectively. We shall also refer B to K_p. As a time scale, we shall use K_s/V, the time it takes the input rate V to achieve a concentration increase of K_s. We thus introduce the new dimensionless variables

$$\alpha = A/K_s, \quad \beta = B/K_p, \quad \gamma = C/K_p, \quad \tau = t/(K_s/V). \tag{2}$$

With these, the system (1) becomes (Exercise 1)

$$d\alpha/d\tau = 1 - \sigma\phi(\alpha, \gamma), \tag{3a}$$

$$d\beta/d\tau = q\sigma\phi(\alpha, \gamma) - k_t\beta, \tag{3b}$$

$$d\gamma/d\tau = k_t h^{-1}\beta - k\gamma, \tag{3c}$$

$$\phi(\alpha, \gamma) \equiv \frac{\alpha(1+\alpha)(1+\gamma)^2}{L + (1+\alpha)^2(1+\gamma)^2}. \tag{3d}$$

1 The amount of phosphodiesterase is reflected in the size of k^*.

The original system (1) depends on eight dimensional parameters. The system (3) contains the six dimensionless parameters

$$\blacklozenge\blacklozenge \quad \sigma = \frac{W}{V}, \quad q = \frac{K_s}{K_p}, \quad k_t = \frac{k_t^* K_s}{V}, \quad k = \frac{k^* K_s}{V}, \quad h, \quad \text{and} \quad L. \tag{4}$$

Particularly important to us will be σ (the ratio of the maximum cAMP production rate to the substrate input rate) and k. The latter, the dimensionless cAMP destruction term, can be thought of as the ratio of the input time scale K_s/V to the hydrolysis time scale $1/k^*$. It is also worth noting at once that the allosteric constant is expected to be large; typically, $L = 10^6$. This means that the unbound enzyme will be almost entirely in the inactive T state, leaving considerable possibility for extracellular cAMP to convert enzyme to the active R state. (For the benefit of those who will refer to the papers of Goldbeter and associates cited earlier, we note that those papers have somewhat different definitions of σ, q, k_t, and k.)

Steady-state solutions

Following the procedure used in several earlier examples, we shall seek steady states and then examine their stability. It is hoped that this will allow us to predict different forms of qualitative behavior that will conform with the four observed types of activity listed earlier.

Denoting steady-state values of the three dimensionless concentrations by $\bar{\alpha}$, $\bar{\beta}$, and $\bar{\gamma}$, we find at once, by setting $\sigma\phi(\bar{\alpha}, \bar{\gamma}) = 1$ [from (3a)] in (3b) and then using (3c), that

$$\bar{\beta} = q/k_t, \quad \bar{\gamma} = q/hk. \tag{5a,b}$$

The steady-state value of dimensionless ATP concentration, $\bar{\alpha}$, is then prescribed by (3a):

$$\phi(\bar{\alpha}, \bar{\gamma}) = \sigma^{-1} \tag{6}$$

Equation (6) is just the quadratic

$$(\sigma - 1)(\bar{\alpha})^2 + (\sigma - 2)\bar{\alpha} - [1 + (1 + \bar{\gamma})^{-2}L] = 0. \tag{7}$$

It is not difficult to show [Exercise 2(a)] that both roots are negative when $\sigma < 1$. In this situation the input rate V is greater than the maximum cAMP synthesis rate W; under such circumstances the substrate ATP will continually accumulate, so that, indeed, no (admissible) steady state is expected. We thus shall assume

$$\sigma > 1. \tag{8}$$

In this case [Exercise 2(b)] there is a single steady state given by (5) and

$$\bar{\alpha} = \frac{-(\sigma-2) + \{(\sigma-2)^2 + 4[1 + L(\bar{\gamma}+1)^{-2}](\sigma-1)\}^{1/2}}{2(\sigma-1)}. \tag{9}$$

Stability of the steady state

To examine the stability of the sole steady-state solution, we proceed in a fashion that is now familiar. We define perturbations, denoted by primes, by

$$\alpha'(t) = \alpha(t) - \bar{\alpha}, \quad \beta'(t) = \beta(t) - \bar{\beta}, \quad \gamma'(t) = \gamma(t) - \bar{\gamma}. \tag{10}$$

Suppose that the equation for $d\alpha/dt$ takes the form

$$d\alpha/dt = F(\alpha, \beta, \gamma). \tag{11}$$

Then, if the perturbations are small, by a straightforward extension of equation (A2.8) the corresponding linearized equation is

$$\frac{d\alpha'}{dt} = \left[\frac{\partial F(\alpha,\beta,\gamma)}{\partial\alpha}\right]_0 \alpha' + \left[\frac{\partial F(\alpha,\beta,\gamma)}{\partial\beta}\right]_0 \beta' + \left[\frac{\partial F(\alpha,\beta,\gamma)}{\partial\gamma}\right]_0 \gamma'. \tag{12}$$

Here the subscript zero indicates that the square brackets are evaluated at the steady state $(\bar{\alpha}, \bar{\beta}, \bar{\gamma})$. In our case, therefore, the stability equations are [Exercise 2(c)]

$$d\alpha'/dt = -\sigma\phi_\alpha \alpha' - (\sigma\phi_\gamma)\gamma',$$

$$d\beta'/dt = (q\sigma\phi_\alpha)\alpha' - k_t\beta' + (q\sigma\phi_\gamma)\gamma', \tag{13}$$

$$d\gamma'/dt = (k_t h^{-1})\beta' - k\gamma',$$

where

$$\phi_\alpha \equiv \frac{\partial\phi(\alpha,\gamma)}{\partial\alpha}\bigg|_{\alpha=\bar{\alpha},\gamma=\bar{\gamma}}, \quad \phi_\gamma = \frac{\partial\phi(\alpha,\gamma)}{\partial\gamma}\bigg|_{\alpha=\bar{\alpha},\gamma=\bar{\gamma}}.$$

Our earlier stability considerations involved systems of two ordinary differential equations; the relevant calculations required careful study of a certain quadratic algebraic equation. In the present case, study of a cubic is required. Later, we shall discuss stability further, but for the present it is sufficient to note that Goldbeter and Segel (1977) analyzed the roots of the appropriate cubic, using a computer, and found that the steady state was stable under some conditions and unstable under other

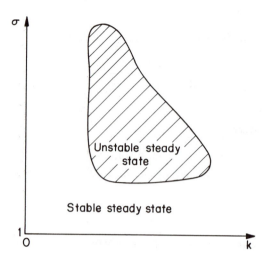

Figure 6.3. Linear stability analysis of the steady state of (5) and (9): Schematic presentation of results for fixed values of all parameters except the cAMP production parameter σ and the destruction parameter k.

conditions. If all parameters except σ and k are kept constant, the stability results can be presented schematically as in Figure 6.3.

Given that there is a single steady state, if this state is found to be stable, then one might conjecture that all solutions approach it. What if the steady state is unstable? Knowing that the *Distyostelium* amoebae can exhibit oscillations in cAMP secretion, we would anticipate that in this case an oscillatory state results, whatever the initial condition. These conjectures are summarized in Figure 6.4.[2]

We have recommended as a general approach to systems of nonlinear equations the following procedure: Find steady states, examine their stability, conjecture qualitative behavior, and check your conjectures with computer simulations. This approach was adopted by Goldbeter and Segel (1977), who used the CSMP computer package (see Preface) to find solutions of the system (3) under various conditions. The conjectures of Figure 6.4 were borne out. Indeed, when parameters were selected so that analysis of (13) predicted a stable steady state, then solutions approached this steady state whatever the initial conditions. By contrast, when the steady state was unstable according to the analysis of (13), then periodic oscillations were found (Figure 6.5).

2 The mathematical term **bifurcation** is used to describe the "splitting off" of a new solution when a parameter passes some critical value. An example is shown in Figure 6.4. For intermediate values of k, when σ exceeds a critical value, the steady state becomes unstable, and an oscillatory solution replaces it as the **attractor** toward which all solutions tend as time passes.

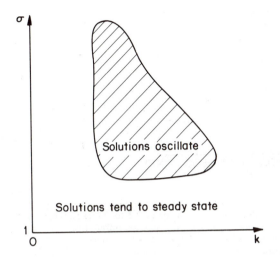

Figure 6.4. Conjectured behavior of the system (3), based on the results of Figure 6.3.

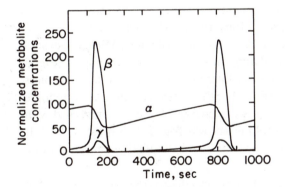

Figure 6.5. Autonomous periodic oscillations of dimensionless concentrations of extracellular cAMP, γ, intracellular cAMP, β, and intracellular ATP, α. Computer solution of (3) with parameter values $V/K_s = 0.1$ sec^{-1}, $W/K_s = 1.2$ sec^{-1}, $k^* = k_t^* = 0.4$ sec^{-1} yielding dimensionless parameters $\sigma = 12$, $k_t = 4$, $k = 4$. Also, $q = 100$, $h = 10$, $L = 10^6$. [From Goldbeter and Segel (1977), Figure 2.]

But what of the relay capability of the cells? Perhaps the cells can amplify an extracellular cAMP signal even though they eventually will return to a steady state. Such an idea is not difficult to check by computer simulation, provided we know which parameter settings to use. [With a six-dimensional parameter space, as in (4), guessing is almost sure to be futile.] We can conjecture that relay is "almost" oscillation; relay may occur under conditions such that for some reason the system

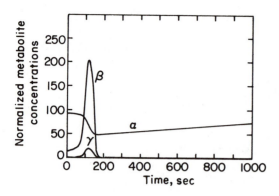

Figure 6.6. Relay of a cAMP signal. Conditions as in Figure 6.5, except that the injection rate V is decreased by a factor of 2.5 (or other suitable changes are made), so that now $\sigma = 30$, $k_t = k = 10$. Initially, intracellular ATP and cAMP have their steady-state values $\bar{\alpha} = 92.4$, $\bar{\beta} = 10$, while extracellular cAMP is set to 2, twice its steady-state value of $\bar{\gamma} = 1$. [From Goldbeter and Segel (1977), Figure 4.]

can provide a single pulse but cannot quite excite itself into successive pulses. This suggests that we should look just outside the border of Figure 6.4 that delineates the parameter domain for periodic solutions. This was done. For initial conditions, the intracellular variables α and β were set at their steady-state values, but extracellular cAMP was elevated above the steady state, simulating an externally applied pulse. Several simulations gave nothing of great interest, but other parameter choices did indeed give relay (Figure 6.6). The dependence of the relay amplitude on the size of the initial signal is depicted in Figure 6.7. In accord with experiment, a distinct threshold effect was found. Subthreshold signals hardly elicit a response, but superthreshold stimulation can yield more than 20-fold amplification.

A rich variety of behavior has now been discovered, as summarized in Figure 6.8. Our mathematical model (3) exhibits all the major states (a)–(d) listed at the beginning of this chapter. Relay and autonomous oscillation have been demonstrated. The steady-state solution (5b) for $\bar{\gamma}$ shows that extracellular cAMP will be negligible when the hydrolysis parameter k is large (right portion of Figure 6.8), whereas small k will be accompanied by a high steady-state concentration of extracellular cAMP.

The developmental path

We now discuss an important conclusion that stems from the observation that the enzyme activities do not in fact remain constant during slime

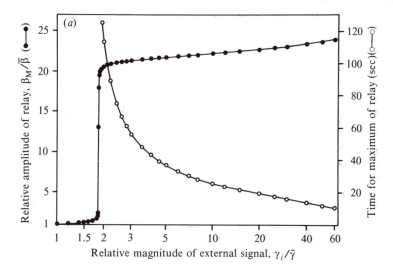

Figure 6.7. Dependence of relay amplitude on the external signal. The abscissa gives the ratio of initial (γ_i) and steady-state ($\bar{\gamma}$) concentrations of extracellular cAMP. The response amplitude is defined as the ratio of the peak intracellular cAMP concentration (β_M) to its steady-state value $\bar{\beta}$. [From Goldbeter and Segel (1977), Figure 5.]

mold development. These activities have been represented by the constant parameters σ and k. This is legitimate because the parameters change significantly in a matter of hours, whereas a cAMP pulse or relay takes only a few minutes. If an enzyme activity changes by 1% or 2% in a few minutes, ignoring such a change should have very little effect on our simulations.[3]

Suppose that slow parameter changes are such that the parameter pair (σ, k) follows the path designated by the heavy arrow in the **parameter plane** of Figure 6.9. According to our model, the cell will "automatically" shift its behavior from seemingly inert, to relay-capable, to oscillating, to strongly and steadily secreting.

Switches in cell activity are generally attributed to the switching on of new sets of genes. Here an alternative is suggested, based on observations that there are slow *developmentally controlled* changes in the maximum activities of the enzymes adenylate cyclase and phosphodiesterase. (The term "developmentally controlled" refers to developmental events

---

3 Consider a point in Figure 6.8 that lies just outside the σ–k domain corresponding to oscillations. Here, very small changes in σ or k *can* result in a large effect on system behavior, by moving the system across the domain boundary. This merely means, however, that we cannot rely on the theoretical boundary as an exact demarcation of the oscillatory domain.

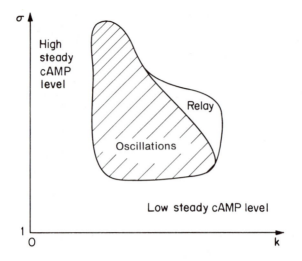

Figure 6.8. Domains of different qualitative behaviors of the model (3) in the parameter plane of cAMP production (σ) and destruction (k).

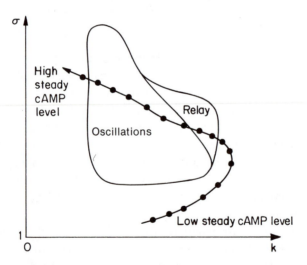

Figure 6.9. Schematic possible developmentally controlled evolution of the enzyme activities σ and k that could account for the observed transitions between states. The heavy dots denote possible parameter values at one-hour intervals, along the "developmental path" in parameter space.

whose onset, duration, and magnitude seem to be essentially the same in each individual of a given species.) The causes of the gradual changes in enzyme activity are not known, but presumably they can be traced back to some slow modification of gene activity.

Previous discussions in this book of how different domains in parameter space correspond to different qualitative behaviors have allowed us to note the possibility that qualitative switches in behavior can arise from slight parameter changes. Thus, the genes may act only indirectly in switching behavior in the *D. discoideum* secretory system; the immediate cause may be slight shifts in enzyme activities that bring about major changes in the operation of the underlying biochemical feedback pathways.

We have now presented the main ideas stemming from the mathematical model of the slime mold signaling system. For those interested in pursuing the problem further, we shall now penetrate into the matter more deeply in two respects. First, we shall take a more detailed look at the model, which will allow us to explain its behavior by means of phase-plane analysis. Second, we shall discuss in more detail the relationship between the model's predictions and experimental observations.

A simplified model

To better understand our numerical results, let us attempt to reduce our system of three equations (3) to two, by means of a suitable quasi-steady-state approximation. But which variable should we suppose is in a quasi-steady state? An answer will be suggested by a study of our numerical simulations. In these computer analyses, parameters were varied so that reasonable agreement with experiment was obtained; we shall take the parameters given in the legend for Figure 6.6 as typical.

The simulation results of Figures 6.5 and 6.6 show that the intense period of activity in both relay and oscillation lasts for about three minutes. In contrast, typical time scales used for cAMP secretion and destruction are $(k_t^*)^{-1} = 2.5$ sec and $(k^*)^{-1} = 2.5$ sec. Thus, both processes that regulate the extracellular cAMP concentration, secretion and hydrolysis, are fast compared with the overall time scale of the phenomena under consideration. This suggests that we assume that γ is in a quasi-steady state. From (3c), this assumption implies that

$$\gamma = k_t \beta / (hk). \tag{14}$$

Indeed, inspection of Figures 6.5 and 6.6 shows that there is a rough proportionality between γ and β, lending further credence to our assumption.

With (14), equations (3a) and (3b) become

$$\frac{d\alpha}{d\tau} = 1 - \sigma\phi\left(\alpha, \frac{k_t \beta}{hk}\right), \quad \frac{d\beta}{d\tau} = q\sigma\phi\left(\alpha, \frac{k_t \beta}{hk}\right) - k_t \beta. \tag{15a,b}$$

We shall make use of one more simplification whose justification comes from the numerical simulations. From Figures 6.5 and 6.6 we see that α always remains large compared with unity. In such circumstances we can replace $\alpha + 1$ in ϕ by α. This gives, approximately,

$$\phi(\alpha, \gamma) = \frac{\alpha^2(1+\gamma)^2}{L+\alpha^2(1+\gamma)^2}. \tag{16}$$

The fact that ϕ can now be regarded as a function of α^2 considerably eases the calculational burden.

Phase-plane analysis of the simplified model

Our simplified problem consists of (15) and (16). We can analyze the phase-plane representation of the solution to these equations by the methods that have already been presented.

We shall first plot the nullclines of our problem in the (β, α) plane. The horizontal nullclines are found by setting $d\alpha/d\tau = 0$. From (15a) and (16) we find that there is one such curve, given by

$$\alpha = \frac{[L/(\sigma-1)]^{1/2}}{1+(k_t/hk)\beta}, \tag{17}$$

or, with the parameter values of Figure 6.5, by

$$\alpha = 302/(1+0.1\beta). \tag{18}$$

This curve is plotted in Figure 6.10.

The calculations required to obtain the vertical nullcline (where $d\beta/d\tau = 0$) are a little more involved, but are not difficult. The details are left to Exercise 3. The result is that the vertical nullcline is given by

$$\alpha = \left(\frac{L\gamma}{(q\sigma/hk)-\gamma}\right)^{1/2}\frac{1}{1+\gamma}, \quad \gamma \equiv \frac{k_t\beta}{hk}, \tag{19a,b}$$

or, with the parameters of Figure 6.5, by

$$\alpha = 10^3\left(\frac{\gamma}{30-\gamma}\right)^{1/2}\frac{1}{1+\gamma}, \quad \gamma = \frac{\beta}{10}. \tag{20a,b}$$

The graph of (19a) has horizontal tangents when $d\alpha/d\beta \sim d\alpha/d\gamma = 0$, that is, when

$$2(hk/q\sigma)\gamma^2 - \gamma + 1 = 0. \tag{21}$$

Because $hk/q\sigma$ is typically small compared with unity, the solutions of (21) are approximately (Exercise 5)

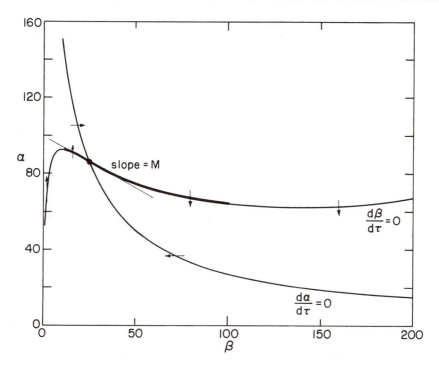

Figure 6.10. The nullclines (18) and (20), corresponding to the oscillatory situation of Figure 6.5. The nullclines intersect on the heavy line, where (27) holds, so that the corresponding steady-state point (heavy dot) is unstable.

$$\gamma = 1, \quad \gamma = q\sigma/(2hk). \tag{22}$$

With the information gathered thus far, we can construct the S-shaped graph of (20a), as is done in Figure 6.10. (The S shape becomes more apparent from later figures, such as Figures 6.12–6.14, where the horizontal scale is contracted.) A little further analysis provides the directions of the vertical and horizontal trajectories that are associated with the nullclines. A few of these are also shown on the figure.

The nullclines intersect in a single point, the sole steady-state solution of the system (15). The linear stability of this solution is governed by the following equations for the perturbations $\alpha' = \alpha - \bar{\alpha}$ and $\beta' = \beta - \bar{\beta}$:

$$\frac{d\alpha'}{d\tau} = a\alpha' + b\beta', \quad \frac{d\beta'}{d\tau} = c\alpha' + d\beta', \tag{23a}$$

where (Exercise 4)

$$\begin{aligned}
a &= -\sigma\phi_\alpha, \quad b = -(\sigma k_t/hk)\phi_\gamma, \\
c &= q\sigma\phi_\alpha, \quad d = (q\sigma k_t/hk)\phi_\gamma - k_t.
\end{aligned} \tag{23b}$$

Here we have employed the abbreviations

$$\phi_\alpha = \frac{\partial\phi(\alpha,\gamma)}{\partial\alpha}\bigg|_{\substack{\alpha=\bar\alpha\\\gamma=\bar\gamma}}, \quad \phi_\gamma = \frac{\partial\phi(\alpha,\gamma)}{\partial\gamma}\bigg|_{\substack{\alpha=\bar\alpha\\\gamma=\bar\gamma}}, \quad \bar\gamma \equiv \frac{k_t\bar\beta}{hk}. \tag{24}$$

Analysis of stability requires study of the two quantities $-(a+d)$ and $ad-bc$ of equation (A5.16). The latter satisfies

$$ad-bc = \sigma k_t \phi_\alpha > 0. \tag{25}$$

The former can be written (Exercise 4)

$$-(a+d) = q\sigma\phi_\alpha[q^{-1}+M], \tag{26}$$

where M is the slope at the steady-state point of the S-shaped vertical nullcline (19a) or (20a) (see Figure 6.10).

Because $ad-bc>0$, the steady state will be stable if and only if $-(a+d)>0$. This means that steady states that are found on the two ascending arms of the S-shaped vertical nullcline are stable, for then $M>0$. Unstable steady states occur when

$$M < -q^{-1}. \tag{27}$$

Because q is typically a large number, steady states located on the descending portion of the vertical nullcline will be unstable, with the exception of the portions near the maximum and the minimum of the S-shaped curve, where slopes are negative but small. The points on the vertical nullcline where (27) holds are designated by a heavy line in Figure 6.10.

Explanation of relay and oscillation

Enough information has already been gathered to allow us to understand better the conditions required for relay. According to our simulations, relay occurs when conditions are "almost" those that result in autonomous oscillations. Furthermore, relay results when the extracellular cAMP concentration γ is increased sufficiently above its steady-state value. By our quasi-steady-state assumption expressed in (14), γ is proportional to the intracellular cAMP concentration β. This means that we must examine the β–α phase plane for conditions in which a suitable increase in β will trigger a pulse of cAMP. Such conditions are found in Figure 6.11, where the steady-state point is located just to the left of the maximum of the S-shaped vertical nullcline. Crossing this nullcline changes the sign of $d\beta/d\tau$. A small increase in β will bring us to a point where $d\beta/d\tau < 0$, resulting in a quick return to steady state. By contrast,

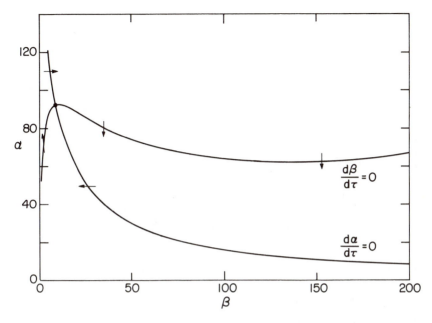

Figure 6.11. Nullclines in a case of relay, corresponding to the situation of Figure 6.6. The steady-state point (heavy dot) is stable.

consider an initial state in which β is increased sufficiently that the initial point lies on the far side of the S curve. Now $d\beta/d\tau > 0$, and β will increase. This increase results in a trajectory that moves rightward until it again crosses a branch of the S curve, so that β begins to decrease again (Figures 6.13 and 6.14).

To understand relay and oscillation in more biological terms, let us first note that $\partial\phi/\partial\alpha > 0$, reflecting the fact that the conversion of substrate ATP into product cAMP is faster if the substrate concentration is increased. Also, $\partial\phi/\partial\gamma > 0$, because product forms faster if more cAMP is bound to the extracellular regulatory site on the catalytic enzyme adenylate cyclase, converting this enzyme into a more active form.

If extracellular cAMP concentration is elevated slightly above its steady value, there is an accompanying increase in the rate of substrate–product conversion. This results in an increase in extracellular cAMP concentration, tending to further increase the rate of cAMP production. In addition, however, the faster product synthesis rate brings about a decrease in substrate concentration, an effect that tends to *decrease* cAMP production. There are conflicting influences here, and their resolution is a quantitative matter that is difficult to predict by pure intuition.

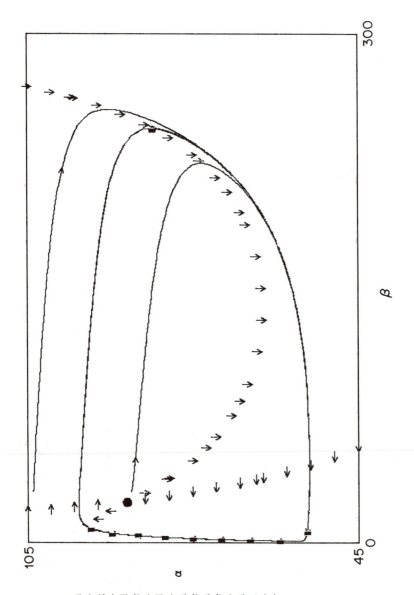

Figure 6.12. Some computer-drawn phase-plane trajectories in the oscillatory situation corresponding to Figure 6.10; $k_I = 4$, $\sigma = 12$. (Figures 6.12–6.14 were constructed by G. Odell using his automatic phase portrait generator.) Nullclines are indicated by vertical and horizontal arrows. The black rectangles are on the limit cycle, a closed curve representing a periodic oscillation. It takes 5.7 sec for a point to travel from one rectangle to the next in the (clockwise) traversal of the limit cycle – giving a total period of 57 sec. As shown, arbitrary solutions generally approach the periodic state rather rapidly.

As we have seen, there are conditions in which a small increase in extracellular cAMP increases the intracellular synthesis rate so much that it can bring about an explosive increase in cAMP production. This increase ceases because the extracellular cAMP binding sites become virtually saturated. In such circumstances catalysis is proceeding at its maximum rate, so that there can be no effect to outweigh the slowing caused by a decreasing substrate level. The reaction thus subsides, and the substrate concentration gradually builds back toward its steady-state value. At this stage, the system is delicately balanced. It may be that it continues to approach a steady state, resulting in a return to quiescence after a cAMP stimulus (relay). But if parameters are just slightly different, the buildup of substrate and the decrease in the control molecule, extracellular cAMP, may bring the system into a state in which the reaction rate starts to build up once again, liberating more extracellular cAMP (γ), further increasing the reaction rate, and so forth. In this state, "spontaneous" periodic cAMP pulses occur. [The periodic solution is represented by a closed curve in the α–β phase plane for the approximate system (15) (see Figure 6.12).]

Phase space and the state point

Let us digress for a moment to call attention to the fact that the behavior of a solution to the governing equations (3) can be regarded as a trajectory $[\dot{\alpha}(t), \beta(t), \gamma(t)]$ in α–β–γ space. This is a three-dimensional phase space, a natural generalization of the two-dimensional phase plane with which we have become familiar. We shall use the terminology **state point** for $[\alpha(t), \beta(t), \gamma(t)]$. The development of the system (3) with time can be visualized as a wandering of the state point on a trajectory in α–β–γ phase space. If many variables were taken into account in our equations, the state of the system would have to be described by the passage of the state point along a trajectory in a correspondingly high-dimensional phase space.

High-dimensional spaces are difficult to visualize, and even trajectories in three-dimensional phase space cannot be represented all that easily in the two-dimensional graphs that appear in papers and books. Thus, there are clear advantages to considering cases in which quasi-steady-state approximations reduce the problem to the study of a state point moving along a trajectory in a phase *plane*.[4] In our case, (14)

---

4 For ordinary differential equations, only in three or more dimensions do certain phenomena occur, such as the continuous analogue of the chaotic solutions described in Chapter 2. Thus, some caution must be exercised in using an approximation that reduces the dimensionality of a problem.

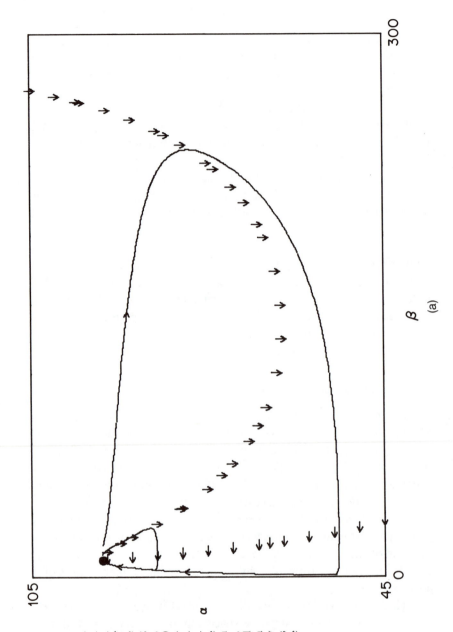

Figure 6.13. (a) Sub-threshold and super-threshold phase-plane trajectories in the case of relay (excitability) in the situation corresponding to Figure 6.11; $k_t = 10$, $\sigma = 30$. (b) Detail of (a) with many more trajectories, showing how a sufficiently large initial displacement of β from the steady state leads to much further β amplification, whereas a small initial displacement results in a direct return to the steady state. See the cover of this book for a colored phase portrait of excitability.

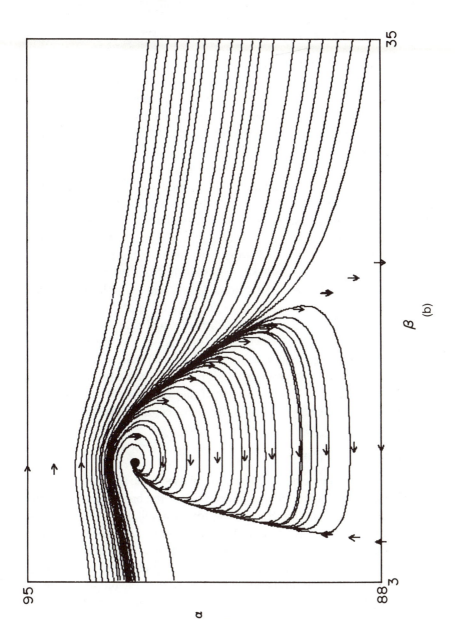

β

(b)

α

95

88 3

35

provides such an approximation, reducing (3) to a system in two dependent variables, $\alpha(t)$ and $\beta(t)$. Let us see how this representation helps us to understand the following phenomenon (see, for example, Gerisch, 1968). If a pulse of extracellular cAMP is provided and relay is induced, then a second pulse will induce a second relay response only if there is a sufficient delay between the pulses.

Imagine that the system is at steady state, but is capable of relay. Rapid addition of extracellular cAMP is modeled by instantaneously moving the state point rightward from its steady-state position, to R in Figure 6.14. Suppose that more cAMP is added when the state point subsequently reaches the position labeled A in Figure 6.14. This has the effect of instantaneously moving the state point farther to the right, but the ensuing trajectory will differ only slightly from the trajectory that represents the undisturbed situation. This is an example of **absolute refractoriness** – a new stimulus does not have an appreciable effect.

On the other hand, compare the effects of equal stimuli at B and at C in Figure 6.14. The first is ineffective, but the second moves the state point across the descending portion of the vertical nullcline, resulting in another burst of cAMP synthesis. This is **relative refractoriness** – sufficiently large stimuli will initiate activity if they are appropriately delayed, and the longer the delay the smaller the required magnitude of stimulus.

Robustness of the results

An important point emerges from our phase-plane analysis, and it is reinforced by our later verbal arguments: Only the broad features of the model, not its details, are responsible for its qualitative behavior. The enzyme adenylate cyclase may not be a dimer, or it may deviate from our model for its behavior in several other particulars. But if the general form of the nullclines is retained, then the same qualitative behavior may emerge from a more accurate representation of how the binding of extracellular cAMP influences the synthesis of intracellular cAMP.

Interaction between analytic and numerical methods

Our discussion provides a good illustration of the interdependence of analytic and numerical methods. We started with some analysis – locating a steady-state point and examining its stability. The analysis delineated regions of differing stability behavior in parameter space. The conjectured differing qualitative behaviors were verified numerically:

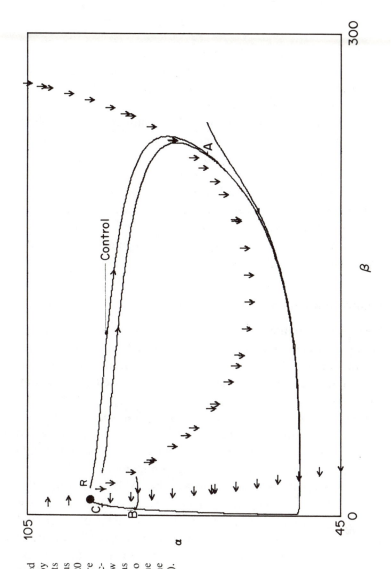

Figure 6.14. Refractoriness demonstrated in the phase plane. Relay is initiated by moving the state point to R. At points A, B, and C there is an instantaneous increase in β, moving the state point 20 units to the right in each case. At A, there is a negligible effect (absolute refractoriness). At B, the stimulus is below threshold, but at C, after more time has elapsed, the same stimulus is sufficient to excite relay (relative refractoriness). The "control" trajectory is the same as the (larger) relay trajectory of Figure 6.13(a).

Oscillations were found for some parameters, and decay to the steady state for others. Because both analytic work and numerical work are prone to error, numerical verification that oscillation actually occurs where predicted provides a gratifying check on our analysis.

In this example we went beyond the by-now-conventional verification of behavior that was conjectured from the results of linear stability analysis. Our computer calculations showed that our model could reproduce another observed aspect of observed behavior: relay capability. Using the results of these calculations as a guide, we were then able to gain further insight by a return to analytic methods. Study of the computer output helped us to select a suitable variable for a quasi-steady-state assumption, and (because α was large compared with unity) it suggested the helpful approximation $\alpha + 1 \approx \alpha$.

Comparison with experiment

We now wish to discuss in some detail the comparison of our theoretical results with experiment. The richness of behavior under investigation and the considerable amount of experimental information available provide a good opportunity to see some of the subtleties involved in such a comparison.

We have seen that our model can explain transitions from inert to relay to oscillatory behavior in cells. For relaying cells, the model correctly predicts thresholds of excitation and the existence of both absolute and relative refractory periods following stimulation.

The availability of sensitive assays for cAMP allows quantitative as well as qualitative comparison. To this end, a trial-and-error procedure, employing repeated computer simulations, led Goldbeter and Segel (1977) to a set of parameter values that give a match with experiment that was mainly as satisfactory as could be expected (Table 6.1). An exception is the delay between peaks of extracellular and intracellular cAMP. Here the theory gave about 3 sec, which is an order of magnitude smaller than the half-minute delay observed by Gerisch and Wick (1975). At first, the only defense that could be made for the discrepancy was the existence of some evidence (Maeda and Gerisch, 1977) that cAMP was packed in vesicles before being secreted, a process that was not taken into account in the model. More recently, however, Dinauer, MacKay, and Devreotes (1980) observed under their experimental conditions that indeed the cAMP secretion rate was directly proportional to the intracellular cAMP concentration, with no significant delay between extracellular and intra-

Table 6.1. *Comparison of experimental and theoretical predictions concerning relay*

	Experiment	Model
Relay amplitude (maximum $cAMP_i$/steady $cAMP_i$)	10–25	20
Time for maximum relay (sec)	100–120	113
Half width (sec)	60	53
Delay between peaks of $cAMP_i$ and $cAMP_e$ (sec)	30–40	3
Minimum ATP/maximum $cAMP_i$	50	25
Maximum $cAMP_i$/maximum $cAMP_e$	20–50	10
Steady state: ATP/$cAMP_i$	1,000	924

Note: Theoretical parameters as in Figure 6.6; i = intracellular; e = extracellular.
Source: Experimental results from Gerisch and co-workers. See Goldbeter and Segel (1977) for references on experiments and for a demonstration that the parameters are in the physiological range.

cellular peaks. Thus, what seemed a deficiency in the theory is perhaps no longer of major concern, in view of more recent experiments, but there remains the question why different experiments show different delays.

Gerisch and associates (1977) found no evidence that the total intracellular ATP concentration varies during relay and oscillation; yet the presence of variation in the ATP concentration α is essential to our model. For parameters essentially the same as those used in the simulations we have reported, however, such variation can be restricted to about 10% of the mean ATP concentration. This is approximately the error of the measurements made by Gerisch and associates (1977). Moreover, these authors assayed the *total* cellular content of ATP. It is quite conceivable that the total concentration of this important chemical does remain virtually unchanged, but that there is significant variation in ATP concentration in the region of interest here – near the membrane.

Crucial to our model are appropriate changes in enzyme activities. Experimental evidence on this point is the demonstration by Klein (1976) that there is almost a 50-fold increase in the activity of adenylate cyclase within two to five hours of starvation.[5] A parallel rise in the phosphodiesterase activity has been measured by Malchow and associates (1972)

5 This may seem like a fast increase in k, contrary to assumption. But as can be seen from Figure 6.15, it is the logarithm of k that governs changes in the system's behavior.

and Klein and Darmon (1977). According to Figure 6.9, the onset of relay and oscillation requires a later fall in phosphodiesterase activity. Such a decrease is indeed observed, owing to the release of an extra-cellular inhibitor (Riedel and Gerisch, 1971), but the effect is not long-lasting, because cAMP pulses repress inhibitor production (Klein and Darmon, 1977; Yeh, Chan, and Coukell, 1978). Moreover, the timing of the inhibitor secretion does not correlate well with the appearance of centers. We conclude that a major qualitative prediction of the model does not conform with experiment, and therefore that the model must be altered.

A modified model

Goldbeter (1980) introduced an appropriate change by adding a single term to the ATP equation (1a), giving (for some constants k_1 and $A_0 \equiv k_1^{-1}V$)

$$\frac{dA}{dt} = V - k_1 A - W\Phi(A, C),$$

or, equivalently,

$$\frac{dA}{dt} = k_1(A_0 - A) - W\Phi(A, C). \tag{28}$$

In its first form, the new term $(-k_1 A)$ can be regarded as taking into account the degradation of ATP from reactions other than those cata-lyzed by adenylate cyclase. In its second form, the new term can be thought of as a shift from a constant ATP input to an input that is pro-portional to the difference between the local concentration A and the essentially fixed concentration A_0 of an intracellular pool. The two effects may coexist.

Figure 6.15 depicts the new form of the various regions in the para-meter plane. This figure is fairly accurate, in contrast to the qualitative results of Figure 6.8. Most important, however, is the fact that the relay region is now placed in such a way that the passive–relay–oscillation transitions can be made in accord with experiment mainly by an increase in the cyclase activity parameter σ, without the necessity for a decrease in the phosphodiesterase activity k.

A new type of behavior appears in the modified model. The region E in Figure 6.15 corresponds to parameter values for which there are three steady states. We shall not discuss region E further here, except to

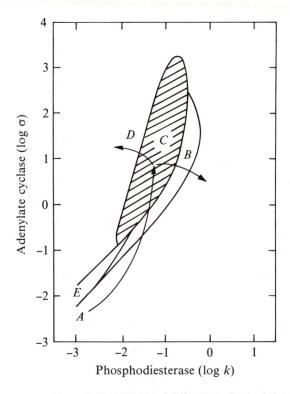

Phosphodiesterase (log k)

Figure 6.15. Domains of different qualitative behaviors of the model (3), with the addition of a decay term $-k'\alpha$ for ATP. From (28), $k'=k_1/V$. Fixed parameter values are $k'=0.025$, $k_t=100$, $q=100$, $h=10$, $L=10^6$. Relay was defined to occur when an initial value of $\alpha=\bar{\alpha}$, $\beta=\bar{\beta}$, $\gamma=10$ produced a subsequent peak for which $\gamma>10$. The arrows represent slow changes of parameters that can shift the cells from passivity (A), to relay capability (B), to oscillatory (C), to a state of high steady secretion (D). [From Goldbeter and Segel (1980), Figure 1.]

remark that Goldbeter (1980) has shown that for certain parameter values, two of these states are stable – which may be relevant to findings of Juliani and Klein (1978) that cAMP pulses can induce a rather long-lasting transition of cyclase from a relatively inactive state to a relatively active state.

In our earlier model, as we have noted, steady states were characterized by relatively high *levels* of extracellular cAMP when the phosphodiesterase activity k was low. Nonetheless, the secretion rate $k_t\beta$ had the constant value q for all values of k and σ. To see this, recall that (3c) implies

$$k_t\bar{\beta}=kh\bar{\gamma}, \tag{29}$$

but all the equations of (3) combine to give

$$kh\bar{\gamma} = q. \tag{30}$$

In the present case, (29) still holds, but now $\bar{\gamma}$ is typically a rather strongly increasing function of the maximum cyclase activity σ (see Figure 4.1.19 of Goldbeter, 1980). Thus, secretion rates will also differ in different portions of the σ–k parameter plane.

Further comparisons with experiment

Here is an example of the sort of prediction that can be generated from the "partitioned parameter plane" of Figure 6.15. Some cells are presumed to pass from the oscillatory region C to region D (high steady cAMP secretion). In the neighborhood of such cells, the phosphodiesterase activity is predicted to decrease. On the other hand, other cells could pass from C back to B. This would occur if there were an additional rise in phosphodiesterase activity – and there is a second wave of phosphodiesterase synthesis that begins five to seven hours after starvation (Klein and Darmon, 1977; Scrive, Guespin-Michel, and Felenbok, 1977). If this increase in k does indeed bring about the transition of cells from the oscillatory state to the excitable state, then this transition should not occur if the second wave of synthesis is abolished. The agent 5-bromodeoxyuridine reduces the second round of synthesis by 80% (Scrive et al., 1977), and the cells remain in the oscillatory regime! (M. Scrive and B. Felenbok, private communication.)

Another use of Figure 6.15 to explain experimental results was made by Newell and Ross (1982). These authors found that adenosine inhibited the initiation of aggregation centers without affecting relay. In one experiment, cells that seemed to be behaving normally in the presence of adenosine were moved from darkness to strong light. This shock brought about the cessation of some oscillations; see frontispiece (top). Careful examination of the picture shows the invasion of featureless territories by waves from adjacent domains – demonstrating that relay capability continues to exist.

Newell and Ross (1982) showed that adenosine strongly inhibited the binding of cAMP to the cAMP receptors on the cell surface. They suggested that the resulting decrease in response of adenylate cyclase to cAMP signals could move the state of the system from domain C to domain B in Figure 6.15 and thereby could bring about the observed shift from oscillating to relay.

The study of mutants is perhaps the most powerful way to challenge biological theories. For example, a mutant with deficient phosphodiesterase production should move approximately vertically from A to D in Figure 6.15. In this case, oscillations should be inducible by adding phosphodiesterase or by decreasing the extracellular cAMP concentration by diffusion. Both of the latter treatments were shown by Darmon, Barra, and Brachet (1978) to induce periodic aggregation in an aggregateless mutant of *Dictyostelium* whose phosphodiesterase production was significantly below normal.

The mutant Fr17 aggregates rapidly because autonomously signaling cells appear much earlier than normal (Durston, 1974). The theory suggests that this may be due to an unusually rapid increase in cyclase activity – and such an increase does occur in Fr17 suspensions (Coukell and Chan, 1980).

Recall that evidence has been cited for the effects of variables on parameters (for example, a tendency for cAMP pulses to repress phosphodiesterase inhibitor and therefore to increase k). A general vista is thereby suggested not only of shifting parameters affecting the qualitative behavior of variables (developmental path) but also of the reverse process. After all, a "parameter" is often merely a variable that changes relatively slowly.

We cannot rule out the possibility that a model quite different from the one we have examined might, in the end, prove to be the correct model. For example, Martiel and Goldbeter (1981) showed that our major results would still hold if activation of adenylate cyclase occurred by a cAMP-dependent protein kinase. In addition, Cohen (1977) presented a system of equations with all the qualitative behaviors of the model studied here, wherein a starvation-induced decrease in a reserve material is assumed to produce appropriate shifts in a catabolite that regulates intracellular cAMP. Cohen originally suggested that the catabolite might be calcium, but there is now evidence (Sussman and Schindler, 1978) that ammonia could play the required role. Further developments of Cohen's model have been discussed by Hagan and Cohen (1981). Of particular interest is their theory for the spiral wave patterns that often occur in aggregating amoebae [see frontispiece (top)].

Relatively recent experiments have made it clear that modification of the model presented here is required. Devreotes and Steck (1979) studied *D. discoideum* cells in conditions in which the extracellular cAMP concentration could be elevated and then held fixed. These authors demonstrated adaptation phenomena that cannot be explained with the earlier models. Furthermore, Dinauer, Steck, and Devreotes (1980) found that when the extracellular cAMP level was held constant, then there was no longer a threshold behavior of the type predicted in Figure 6.7. To account for these findings, Goldbeter and Martiel (1982) proposed a model that takes into account receptor desensitization, evidence for which has been reported (in *D. discoideum*) by Klein (1979).

We pointed out earlier that our results have a robust character because the qualitative behavior relies heavily on certain broad features of the model, chief among which is the S-shaped nullcline. All of the models that we have mentioned have this feature.

If only the broad features of a model are required to produce a variety of qualitative results, then the model is robust, but on the other hand it is very difficult to pin down precisely what chemical interactions do in fact govern the system. Thus, it is quite possible that additional important interactions must be taken into account before even the cAMP signaling part of aggregation and morphogenesis in *D. discoideum* is fully understood. Perhaps some combination of the models we have mentioned will be found to operate. Perhaps other possibilities must be considered, such as cyclase regulation by the synthesis of an activator and (more slowly) an inhibitor on binding of cAMP to its receptor (Devreotes and Steck, 1979; Dinauer et al., 1980).

Whatever the ultimate answer, it is already remarkable how much relevant behavior can be deduced from such a simple model. The model's basic components of enzyme cooperativity and feedback through the secretion of regulator molecules are so widespread that it would be surprising if cells did not evolve to make use of the possibilities that are inherent in the interaction of such components.

Exercises[6]

1. Verify that introduction of the variables (2) transforms (1) into (3). check that all the parameters of (4) are dimensionless.
2. (a) Verify equation (7), and show that both its roots are negative when $\sigma < 1$.
   (b) Show that (7) has a single positive root, (9), when $\sigma > 1$.
   (c) Verify (13).
3. Carry out the detailed calculations necessary to verify that Figure 6.10 correctly gives the shapes of the nullclines and the directions of the trajectories thereon. Also show that trajectories head inward across the axes, and explain why this implies that the concentrations will stay positive.

---

6 Also see Exercise 9.4.

4. Verify equations (23b) and (26).
5. Consider the quadratic equation for γ:

$$\epsilon\gamma^2 - \gamma + 1 = 0, \quad 0 < \epsilon \ll 1.$$

(a) By approximating the square root in the quadratic formula, show that the two roots γ_1 and γ_2 of the quadratic satisfy $\gamma_1 \approx 1$, $\gamma_2 \approx \epsilon^{-1}$.
(b) Show that the assumption that $\epsilon\gamma^2$ is negligible and $\gamma \approx 1$ is a consistent way of looking at the quadratic. Demonstrate that the assumption that $-\gamma$ is negligible, while $\epsilon\gamma^2 \approx -1$, is not consistent. Show finally that the third possible matching of terms is consistent and yields the correct first approximation to γ_2.
(c) Use reasoning of the type in part (b) to show that approximations to the three roots of $\epsilon\gamma^3 + \gamma^2 - 1 = 0$ are $\gamma = 1$, $\gamma = -1$, and $\gamma = -\epsilon^{-1}$. [For further development of these ideas, see Lin and Segel (1974, Section 9.1).]

7

Diffusion

In this chapter, the equation that describes the macroscopic process of diffusion is introduced and applied to some biological situations.[1] Chapter 8 makes further use of the material presented here.

Diffusion is the spreading of particles, ranging from molecules to bacteria, whose individual trajectories can be regarded as random. A typical example of diffusion is the dispersion of dye particles that are released into a clear fluid from a hypodermic needle. (If the spread is to be purely diffusive, then release must be gentle, to avoid a "push" in a preferred direction.)

By its very nature, diffusion involves concentrations that change in space. This means that we must consider concentration functions C that depend on space as well as time. In the equation governing diffusion, partial derivatives of C with respect to the spatial variables will be seen to be related to the partial derivative of C with respect to time. Such an equation involving partial derivatives is an example of a **partial differential equation**, the chief mathematical novelty of this chapter. Several aspects of integral calculus are also required for the first time.

We first discuss the derivation of the diffusion equation. This equation will then be used as the basis of a way to ascribe a quantitative measure to bacterial motility.[2] Finally, we treat the important "rule of thumb" for the expected distance that particles diffuse in a given time.

The general balance law

We now derive a general balance law that constitutes the framework of most partial differential equations used in mathematical biology, or indeed in the natural sciences generally. We shall then postulate a partic-

---

1 The derivation of the diffusion equation is quite similar to the author's earlier treatment of this topic (Segel, 1980, Section 6.1). Several paragraphs and Figures 7.1–7.3 are reproduced from that source with little or no change.

2 "Motility" will be used to mean "amount of movement." If creatures can move, they are "motile," in contrast to self-propelled "mobile" machines.

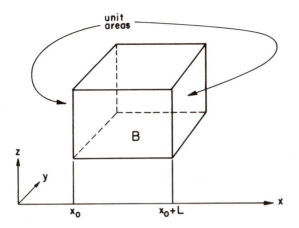

Figure 7.1. A rectangular parallelepiped B. [From Segel (1980), Figure 6.1.1.]

ular form of flow behavior to supplement the general balance law; this leads to the diffusion equation.

For simplicity, let us restrict ourselves to a situation in which all variation is with respect to the single spatial coordinate x. Thus, we shall define $C(x, t)$ as the particle density or particle concentration at a point x and time t. This means that C gives, at time t, the number of particles per unit volume in a small region centered at x. Recall from Figure 5.2 and the accompanying text the nature of the approximation we make in assuming, as we do, that C is a smooth function of t. We also assume that C varies smoothly with x.

Consider the number of particles located within a rectangular box B bounded by unit areas in the planes $x = x_0$ and $x = x_0 + L$ (Figure 7.1). We hypothesize the **general balance law**, which states, in words, that the rate of change of particles in B equals the net creation rate of particles in B plus the net rate at which particles flow into B across its boundaries.

We now make some definitions that will enable us to write the balance law in mathematical terms.

Let $Q(x, t)$ be the **net creation rate**, the birthrate per unit volume minus the corresponding death rate. (For chemical molecules, "births" arise from the combination of other molecules, and "deaths" from the breakup of the given molecular species.) Also, let the **flux density** or **current** $J(b, t)$ give the net rate at which particles cross a unit area in the plane $x = b$ (where b is any number). A net crossing rate is counted as positive (negative) if a majority of particles are crossing in the direction of increasing (decreasing) x (Figure 7.2).

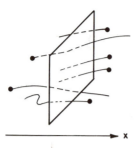

Figure 7.2. Particle trajectories passing through a unit area that is perpendicular to the *x* axis, during a certain unit of time. The flux density for this period is $4 - 2 = 2$ particles per unit area per unit time. [From Segel (1980), Figure 6.1.2.]

With the aid of the foregoing definitions, the general balance law takes the mathematical form

$$\frac{\partial}{\partial t} \int_{x_0}^{x_0+L} C(x, t)\, dx = \int_{x_0}^{x_0+L} Q(x, t)\, dx + J(x_0, t) - J(x_0 + L, t), \qquad (1)$$

for any constants x_0 and L. Note that the signs on the J terms have been taken in accordance with the necessity of keeping track of the rate at which particles pass *into B*. Note also that we have counted only the particle flow through the boundaries marked "unit areas" in Figure 7.1. Because of our assumption that there is no variation with z and y, then during any time interval exactly as many particles flow into B through one of the other two pairs of faces as flow out through the opposite face.

To put the balance law into more usable form we employ the integral mean-value theorem (A8.10) to write (1) as follows:

$$\frac{\partial}{\partial t} C(q_1, t)L = Q(q_2, t)L + J(x_0, t) - J(x_0 + L, t). \qquad (2)$$

Here, all we know of q_1 and q_2 is that

$$x_0 \leqslant q_1 \leqslant x_0 + L, \quad x_0 \leqslant q_2 \leqslant x_0 + L.$$

Equation (2) is true for arbitrary positive L. We choose to divide by L and consider the limit $L \to 0$. In this limit, the unknown points q_1 and q_2 are forced closer and closer to x_0, so that (assuming that C is continuous) we obtain the result

$$\frac{\partial C(x_0, t)}{\partial t} = Q(x_0, t) - \frac{\partial J(x_0, t)}{\partial x}. \qquad (3)$$

Because x_0 is arbitrary, we can write the general balance law as the differential equation

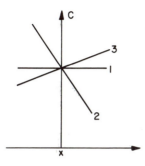

Figure 7.3. Three possible forms, near point x, of a graph of concentration C versus distance. [From Segel (1980), Figure 6.1.3.]

♦♦
$$\frac{\partial C(x,t)}{\partial t} = Q(x,t) - \frac{\partial J(x,t)}{\partial x}. \tag{4}$$

Fick's law

To proceed further, we shall temporarily confine our attention to inert particles that are neither created nor destroyed, so that $Q = 0$. Let us attempt to formulate the simplest reasonable law that might describe the diffusion of these particles. What we must do is relate $J(x,t)$ to the configuration of particles in the neighborhood of x. But concentrations near x can be obtained by means of Taylor series if we know all partial derivatives of C at x. (The present discussion is for a fixed time t, so that time can essentially be ignored for the moment.) Indeed, if we make the notational changes $x - a \to h$, $f \to C$ in the Taylor-series formula (A2.4), we obtain

$$C(x+h,t) = C(x,t) + h\frac{\partial C(x,t)}{\partial x} + \frac{h^2}{2!}\frac{\partial^2 C(x,t)}{\partial x^2} + \frac{h^3}{3!}\frac{\partial^3 C(x,t)}{\partial x^3} + \cdots . \tag{5}$$

This shows explicitly how to calculate all concentrations sufficiently near x (h must be small enough for the series to converge) in terms of the partial derivatives at x. Because of this possibility, we can translate the reasonable assumption that "the flux $J(x,t)$ depends on concentrations near x" into the equation

$$J(x,t) = F\left[C(x,t), \frac{\partial C(x,t)}{\partial x}, \frac{\partial^2 C(x,t)}{\partial x^2}, \cdots \right]. \tag{6}$$

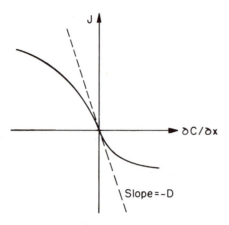

Figure 7.4. Typical graph of flux J as a function of the concentration gradient $\partial C/\partial x$. The dashed line provides an approximation to the graph that is accurate when $\partial C/\partial x$ is sufficiently small.

To progress, consider the expected flux if the graph of C near x has the three different possible forms illustrated in Figure 7.3. In case 1, the flux density J is expected to be zero, for in the course of their random motion, just as many particles will cross from left to right as will cross from right to left. In case 2, J is expected to be positive, for there are more molecules to the left of x, and hence more are expected to pass from left to right than from right to left.

Correspondingly, J is expected to be negative in case 3. This discussion suggests that we might be able to obtain an adequate characterization of the flux if we delete all but the arguments C and $\partial C/\partial x$ from the function F in (6). We have seen that eliminating $\partial C/\partial x$ as well would be too drastic a simplification; knowing only C, we could not even guess the direction of flux.

If we measured J for a number of different values of $\partial C/\partial x$, for fixed C, we might expect a result of the type depicted in Figure 7.4. To simplify matters further, let us now consider a class of experiments in which $\partial C/\partial x$ is sufficiently small to permit approximation of the curve in Figure 7.4 by a straight line through the origin with negative slope. That is,

$$J = -D\frac{\partial C}{\partial x},\qquad(7)$$

where the **diffusivity** or **diffusion coefficient** $D = D(C)$ is a positive quantity that in general will be different for different concentrations. This simplest reasonable assumption for the dependence of flux on concentration is called **Fick's law**. If we introduce it and the assumption $Q = 0$ into the general one-dimensional balance law (4), we obtain the one-dimensional diffusion equation

$$\frac{\partial C}{\partial t} = \frac{\partial}{\partial x}\left[D(C)\frac{\partial C}{\partial x}\right]. \tag{8}$$

In practice, concentration variations often are not sufficient to produce noticeable deviations in D (i.e., D can be regarded as a constant). In this case the diffusion equation takes the more usual form

$$\frac{\partial C}{\partial t} = D\frac{\partial^2 C}{\partial x^2}. \tag{9}$$

The three-dimensional generalization of (9) is

$$\frac{\partial C}{\partial t} = D\left[\frac{\partial^2 C}{\partial x^2} + \frac{\partial^2 C}{\partial y^2} + \frac{\partial^2 C}{\partial z^2}\right]. \tag{10}$$

We have derived (9) and (10) as the simplest reasonable overall description of random particle motion, but decades of use have shown these equations to be suitable in a very wide variety of situations (Crank, 1975).

A typical manifestation of diffusion is the gradual spreading of a dye drop that is released in a colorless fluid. Each dye molecule performs an erratic movement that appears random and hence unpredictable, but experiments have shown that the collective dye movement is well predicted by the appropriate solution of the diffusion equation. In particular, if only the diffusivity D is somehow known, then the entire overall behavior of the dye particles is determined.

Diffusivity of motile bacteria

By analogy with the foregoing discussion, Segel, Chet, and Henis (1977) argued that the random motion of motile bacteria (or other motile organisms), like the random motion of molecules, can be characterized by a single parameter, an effective diffusivity. Such a procedure can replace earlier descriptive and somewhat subjective terms ("very motile," "poorly motile") by an objective estimate of the diffusivity parameter.

Segel and associates (1977) suggested that a convenient quantitative motility assay can be obtained by placing a capillary tube into a suspension of bacteria, waiting for T minutes, and then counting the bacteria in the tube using any of several standard techniques. By comparing the results with theory, one can estimate the bacterial diffusivity D.

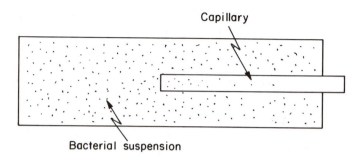

Figure 7.5. Schematic side view of an experimental setup for the motility assay. A capillary tube is inserted into a suspension of bacteria of known concentration, and the bacteria that enter the tube during a prescribed time period are counted.

The experimental setup is diagrammed in Figure 7.5. A mathematical model of the situation requires solution of the diffusion equation (10). The equation must be supplemented by **boundary conditions** which assert (in mathematical terms) that bacteria cannot pass through the boundaries of the fluid in which they are immersed. In addition, one must prescribe **initial conditions** which assert that when the experiments start, bacteria are uniformly distributed outside of the tube and are entirely absent from its interior. Boundary and initial conditions generally are required supplements to partial differential equations that govern changes in space and time. This results in a rather formidable mathematical problem, whose solution would require quite a large computing effort.

There seems to be no biological justification for a highly accurate determination of the bacterial diffusivity – and even if there were, factors such as biological variability and the expected errors in counting would render a heroic effort to obtain an "exact" solution of doubtful value. It thus appears sensible to approximate the original mathematical model. What we lose in (unneeded) accuracy we gain in simplicity.

The first major simplification is to assume that the tube is infinitely long. At first, this may not seem helpful, but it is advantageous not to have to calculate the effect of the closed end of the tube on the flow of bacteria. And such an effect should be negligible if we stop the experiment before more than a few bacteria have managed to swim all the way to the end of the tube. Similarly, if the tube radius is small compared with the width of the chamber, the walls of the latter can be assumed to be infinitely distant.

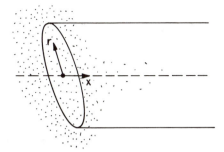

Figure 7.6. The mouth of the tube in the capillary assay, shortly after bacteria have been permitted to enter the tube. The bacterial concentration varies with the axial coordinate x and also with respect to radial distance r.

At least in an idealized experiment, the initial concentration of bacteria just outside the mouth of the tube is the given value C_0, and the initial concentration is zero throughout the tube interior. After a few seconds, the concentration at the mouth will have dropped below C_0, and some bacteria will have entered the tube. A point near the tube boundary is less accessible than a point in the middle, so that the bacterial concentration should be higher toward the middle than toward the edge (Figure 7.6). This leads to a variation in bacterial concentration as a function of the tube radius, in addition to the variation with axial distance down the tube.

The radial variation of concentration is a detail that seems negligible in a first analysis of the problem. Indeed, it appears reasonable to ignore the details at the tube mouth altogether and merely to assume that the mouth concentration is always C_0. This means that the bacterial concentration C in the tube will depend only on the distance x down the tube. The following mathematical problem for $C(x, t)$ emerges (Figure 7.7):

(a) The randomly moving bacterial concentration is governed by the diffusion equation.

$$\frac{\partial C}{\partial t} = D \frac{\partial^2 C}{\partial x^2}, \quad 0 < x < \infty, \quad t > 0. \tag{11a}$$

(b) At the mouth of the tube, the concentration is always C_0.

$$C(0, t) = C_0, \quad t > 0. \tag{11b}$$

(c) There are no bacteria in the tube at the beginning of the experiment.

$$C(x, 0) = 0, \quad x > 0. \tag{11c}$$

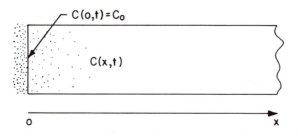

Figure 7.7. Simplified model of the capillary assay, in which it is assumed that the tube is semiinfinite and that the bacterial concentration at the mouth of the tube retains a fixed value C_0.

Most biologists probably would not wish to devote themselves to the study of mathematics long enough to be able to solve with confidence problems in partial differential equations such as (11). But even with the knowledge gained thus far, the reader often can verify rather easily that a proposed solution does in fact solve a given problem in differential equations. In the present case, we can find the solution to (11) in classic treatments of diffusion (e.g., Crank, 1975, section 2.42):

$$C(x,t) = 2C_0 \left[1 - \frac{1}{\sqrt{(2\pi)}} \int_{-\infty}^{z} \exp(-s^2/2)\, ds \right],$$

$$z \equiv x(2Dt)^{-1/2}, \quad t > 0. \tag{12}$$

To verify that (12) satisfies the differential equation (11a), we use the chain rule (A1.10) and the fundamental theorem of calculus (A8.8) to carry out the following calculations (which the reader should carefully check):

$$\frac{\partial C(x,t)}{\partial t} = -\frac{2C_0}{\sqrt{(2\pi)}} \frac{d}{dz} \int_{-\infty}^{z} \exp(-s^2/2)\, ds \Bigg|_{z=x(2Dt)^{-1/2}} \frac{\partial z}{\partial t}$$

$$= -\frac{2C_0}{\sqrt{(2\pi)}} \cdot \exp(-z^2/2) \Bigg|_{z=x(2Dt)^{-1/2}} \cdot x \cdot -\frac{1}{2}(2Dt)^{-3/2} \cdot 2D$$

$$= \frac{2DC_0 x}{\sqrt{(2\pi)}}(2Dt)^{-3/2} \exp(-x^2/4Dt). \tag{13}$$

In like manner,

$$\frac{\partial C(x,t)}{\partial x} = -\frac{2C_0}{\sqrt{(2\pi)}} \frac{d}{dz} \left[\int_{-\infty}^{z} \exp(-s^2/2)\, ds \right]_{z=x(2Dt)^{-1/2}} \frac{\partial z}{\partial x}$$

$$= -\frac{2C_0}{\sqrt{(2\pi)}} \exp(-x^2/4Dt) \cdot (2Dt)^{-1/2}.$$

Thus,

$$D\frac{\partial^2 C(x,t)}{\partial x^2} = -\frac{2DC_0 0}{\sqrt{(2\pi)}} \cdot (2Dt)^{-1/2} \exp(-x^2/4Dt)\frac{\partial}{\partial x}\left[-\frac{x^2}{4Dt}\right]$$

$$= \frac{2DC_0 x}{\sqrt{(2\pi)}}(2Dt)^{-3/2}\exp(-x^2/4Dt) = \frac{\partial C(x,t)}{\partial t}. \tag{14}$$

Verification that (12) satisfies the additional conditions (11b) and (11c) is discussed in the supplement to this chapter.

We now wish to use the solution (12) to calculate the number of bacteria N that are expected to enter the tube in the time interval from $t = 0$ to $t = T$. If the cross-sectional area of the tube is A, then the total number of bacteria in the tube at time T is

$$N = A\int_0^\infty C(x,T)\,dx. \tag{15}$$

[Compare (A8.2).] This integral can be evaluated exactly (see the supplement to this chapter), with the result

$$N = 2C_0 A(TD/\pi)^{1/2}. \tag{16}$$

In the suggested assay, it is desired to estimate the diffusion coefficient D by counting the number N of bacteria that enter the tube in a specified time T. This is readily accomplished by solving (16) for D:

◆◆
$$D = \frac{\pi N^2}{4C_0^2 A^2 T}. \tag{17}$$

Of the quantities that appear in (17), only the diffusion coefficient D is an intrinsic property of the bacteria. At the disposal of the experimenter are the concentration C_0 of bacteria in the suspension, the cross-sectional area A of the capillary, and the duration T of the experiment. If the theory is correct, the same value of D should be obtained (for the same organisms) whatever the choices of C_0, A, and T.

Segel and associates (1977) reported an experimental test of (17) in which C_0 was 7×10^7 ml^{-1}, and the experiment was run for $T = 2, 5, 10,$ 12.5, 15, and 20 minutes, with respective counts N of 1,800, 3,700, 4,800, 5,500, 6,700, and 8,000 bacteria in a capillary of length 32 mm, capacity 1 μl. In addition, counts were obtained of 1,350, 2,300, 3,400, and 6,200 bacteria, for $T = 10$ minutes, with concentrations of 2.5, 4.6, 5.0, and 12.0×10^7 bacteria per milliliter. All these points are plotted in Figure 7.8. It is seen that a value $D = 0.2$ cm^2/hour gives an adequate description of the data.

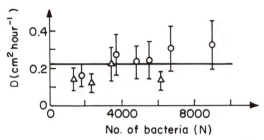

Figure 7.8. Experimental values of the effective bacterial diffusion constant D, calculated according to (17), from experiments described in the text. Circles represent data taken at various times T. Triangles represent data appropriate to different bacterial concentrations C_0. [From Segel et al. (1977), Figure 1.]

Refinement of the experimental technique and of the theory could lead to closer agreement between theory and experiment. As it stands, however, the assay based on (17) seems to provide a fairly easy and reliable way to assign a quantitative estimate D to bacterial motility.

The rule of thumb for diffusion

We now discuss an important general conclusion that is illustrated by the solution to the problem (11). Recall that this problem concerns the diffusion into a semiinfinite tube of a substance that is held at a fixed concentration C_0 at the entrance of the tube. Initially the tube is completely free of the substance in question. The solution (12) can be written

$$\frac{C(x,t)}{C_0} = F(z), \quad z = \frac{x}{\sqrt{(2Dt)}}, \tag{18}$$

where

$$F(z) = 2\left[1 - \frac{1}{\sqrt{(2\pi)}} \int_{-\infty}^{z} \exp(-s^2/2)\, ds\right]. \tag{19}$$

Figure 7.9 depicts a graph of $F(z)$.

We can read from the graph that the locus of points where C is half the entrance value (i.e., where $C/C_0 = 0.5$) is approximately given by $z = \frac{2}{3}$ [i.e., by $x \approx \sqrt{(Dt)}$]. Moreover, $C/C_0 = 0.1$ when $z \approx \frac{5}{3}$ [i.e., $x \approx 2\frac{1}{3}\sqrt{(Dt)}$]. Whether the depth of penetration of the diffusing substance is characterized as the place where it is 50%, 10%, or some other reasonable value of C_0, it appears that in time T, diffusing particles travel a distance L, whose order of magnitude is given by

♦♦ $$L \approx \sqrt{(DT)}. \tag{20}$$

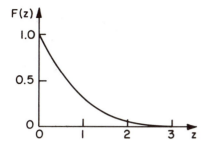

Figure 7.9. Graph of $F(z)$, defined in (19).

This **rule of thumb for diffusion** has been confirmed by solutions of many different types of diffusion problems (Crank, 1975).

As an application of the rule of thumb, let us consider a consequence of the fact that in a typical biochemical reaction the reacting molecules remain bound together for a time of order 1 msec. Small molecules (molecular weight approximately 100) have a diffusion coefficient of approximately 10^{-5} cm^2/sec. Such molecules can traverse a bacterial cell (a distance of about $L = 1\ \mu\text{m} = 10^{-4}$ cm) in approximately

$$T = L^2/D = (10^{-8}\ \text{cm}^2)/(10^{-5}\ \text{cm}^2\ \text{sec}^{-1}) = 10^{-3}\ \text{sec} = 1\ \text{msec}. \qquad (21)$$

Reactions in bacteria will therefore not be unduly slowed because of a necessity to wait for the next required molecule to arrive from somewhere in the cell. For a less primitive eucaryotic cell, however, $L \approx 10\ \mu\text{m}$, so that it will take a molecule about 100 msec to traverse a cell diameter. This simple calculation indicates that some method must be found in eucaryotic cells to keep reactants concentrated, and indeed there are many organelles that have this function. See S. Hardt's Section 6.2 in Segel (1980) for other examples of reasoning of this type.

Exercises

† 1. Adler (1969) found that about 3,000 *E. coli* K12 bacteria entered a capillary tube in one hour, under the conditions of the assay described in the text. It was also reported that 11,000 bacteria of a different strain entered the same tube in 90 minutes, but otherwise under the same experimental conditions. Is the extra time sufficient to explain the extra number of bacteria that entered, or do we also need to conclude that the second strain possessed significantly greater motility than the first?

2. Show that

$$C(x,t) = C_0 \left[1 - \frac{4}{\pi} \sin\left(\frac{\pi x}{L}\right) \exp(-\pi^2 Dt/L^2) \right]$$

is a solution of the diffusion equation (9) appropriate to $0<x<L$, with insulating walls at $x=0$ and $x=L$. (Here, L, C_0, and D are constants.) What initial condition is satisfied by this solution?

3. Many biological situations result in traveling waves of some type. The simplest wave equation, for a function $u(x,t)$, is the partial differential equation

$$\frac{\partial^2 u}{\partial t^2}=c^2\frac{\partial^2 u}{\partial x^2}. \tag{22}$$

† (a) Show that this equation has a solution of the form

$$u(x,t)=f(z), \quad \text{where } z=x-ct,$$

for any function f (which has a second derivative).

(b) Show that $f(x-ct)$ truly represents a traveling wave of speed c, by choosing an arbitrary function $f(x)$ and then sketching $f(x)$, $f(x-c)$, $f(x-2c)$, and so forth.

(c) Show that equation (22) has a solution of the form $g(x+ct)$ for any twice-differentiable function g. How does this solution differ from $f(x-ct)$?

(d) Show that equation (22) possesses the properties of linearity (Appendix 4), in that if $\alpha(x,t)$ and $\beta(x,t)$ are solutions, then so is $\alpha(x,t)+\beta(x,t)$ and $C\alpha(x,t)$ for any constant C.

† (e) It can be shown that the most general solution to (22) is

$$u=f(x-ct)+g(x+ct)$$

for arbitrary twice-differentiable functions f and g. Choose f and g to satisfy the initial conditions

$$u(x,0)=F(x), \quad F\text{ a given function}; \quad \frac{\partial u}{\partial t}(x,0)=0.$$

† (f) Sketch the solution for several times t, given that $F(x)=0$, $|x|\geqslant1$; $F(x)=(1-x^2)^4$, $|x|\leqslant1$. Describe the result.

4. The linear equation (22) is appropriate for various wave phenomena in physics, but biological waves (for example, voltage pulses that travel along nerves) typically are associated with one or more nonlinear reaction-diffusion equations. A prototype equation of this kind for $u^*(x^*,t^*)$ is

$$\frac{\partial u^*}{\partial t^*}=Au^*(B-u^*)+D\frac{\partial^2 u^*}{\partial(x^*)^2}. \tag{23}$$

(a) Show by a suitable choice of dimensionless variables u, x, and t that (23) can be put in the form

$$\frac{\partial u}{\partial t}=u(1-u)+\frac{\partial^2 u}{\partial x^2}.$$

(b) Assume that $u=f(z)$, where $z=x-ct$. Find the ordinary differential equation satisfied by f. [This equation can now be treated by phase-plane methods, because it can be written in the form

$$\frac{df}{dz}=g, \quad \frac{dg}{dz}=-cg-f(1-f).$$

See G. Odell's Section 6.7 and Appendix A.3 in Segel (1980)].

5. Keller and Segel (1971) gave reasons for adopting the following equations as a model for experiments by J. Adler in which a cloud of motile bacteria [density $b(x,t)$] moved steadily down a liquid-filled capillary tube, consuming nutrient [density $s(x,t)$] as they swam along:

$$\frac{\partial s}{\partial t} = -kb, \quad \frac{\partial b}{\partial t} = \mu\frac{\partial^2 b}{\partial x^2} - \chi\frac{\partial}{\partial x}\left(b\frac{\partial s}{\partial x}\right).$$

Here, in addition to a random component, there is a tendency for the bacteria to move toward relatively high concentrations of nutrient. Diffusion of nutrient is neglected. The quantities k and μ are taken to be constants, and reasons are given for assuming that

$\chi = \delta s^{-1}$, δ a constant.

(a) If $b(x,t) = B(z)$, $s(x,t) = S(z)$, $z = x - ct$, and c is a constant, find the equations for B and S.
(b) Keller and Segel (1971) asserted that the equations of part (a) have the solutions

$$\frac{S(z)}{s_\infty} = (1+e^{-\bar{z}})^{-1/(\bar{\delta}-1)}, \quad \frac{B(z)}{c^2 s_\infty(\mu k)^{-1}} = \frac{1}{\bar{\delta}-1}e^{-\bar{z}}(1+e^{-\bar{z}})^{-\bar{\delta}/(\bar{\delta}-1)}, \tag{24}$$

where $\bar{\delta} = \delta/\mu$, $\bar{z} = cz/\mu$. Perform the lengthy calculations required to verify this assertion.
(c) Show from (24) that if $\delta > \mu$,

$$\lim_{z\to\infty} s = s_\infty, \quad \lim_{z\to\infty} b = \lim_{z\to-\infty} s = \lim_{z\to-\infty} b = 0.$$

Show that S continually increases, whereas B has a single maximum. Sketch the graphs of these functions. What is the biological meaning of s_∞?
(d) Integrate the equation for dS/dz from $-\infty$ to ∞, and thereby derive the relation

$$c = Nk/(as_\infty), \tag{25}$$

where N is the total number of bacteria in the tube (assumed infinitely long) and a is its cross-sectional area. Is there a reasonable dependence of c on the various parameters in (25)?
(e) "There are no infinitely long tubes; so the whole analysis is ridiculous." Comment.

Supplement: Verification of equation (7.12)

We shall verify that (7.12) yields (7.11b), (7.11c), and (7.16). In an attempt to verify (7.11b), we set $x=0$ in (7.12), obtaining

$$C(0,t) = 2C_0\left[1 - \frac{1}{\sqrt{(2\pi)}}\int_{-\infty}^{0} \exp(-s^2/2)\,ds\right]. \tag{1}$$

The reader will be familiar with the normal distribution

$$f(s) = \frac{1}{\sqrt{(2\pi)}}\exp(-s^2/2), \tag{2}$$

which (like all probability distribution functions) satisfies

$$\int_{-\infty}^{\infty} f(s)\, ds = 1. \tag{3}$$

From the fact that f is symmetric about the origin $[f(s)=f(-s)]$ it follows that

$$\int_{-\infty}^{0} f(s)\, ds = \int_{0}^{\infty} f(s)\, ds = \frac{1}{2}. \tag{4}$$

The desired result $C(0,t)=C_0$ now follows at once from (1).

Verification of (7.11c) requires us to note that for any positive x, $z \to +\infty$ as t decreases to zero through positive values. [What happens for "negative time," before the experiment starts, is irrelevant. And, strictly speaking, our solution does not satisfy (7.11c) itself but a limiting version of it.]

$$C(x,0) = 2C_0 \left[1 - \frac{1}{\sqrt{(2\pi)}} \int_{-\infty}^{\infty} \exp(-s^2/2)\, ds \right] = 0, \tag{5}$$

where we have employed (3).

More complete treatment of the model of Figure 7.7 would require explicit imposition of the condition that no bacteria flow through the walls of the tube. For a circular tube, we must be concerned with the radial flux, which is given by $J_{\text{radial}} = -D(\partial C/\partial r)$, in analogy with (7.7). This would have to vanish at the boundary of the tube. But because there is no radial variation whatever in our model, $\partial C/\partial r \equiv 0$, and the condition is automatically satisfied.

To verify (7.16) we must calculate the integral in (7.15). From (7.12), this is

$$2C_0 A \int_{0}^{\infty} \left[1 - \frac{1}{\sqrt{(2\pi)}} \int_{-\infty}^{x(2DT)^{-1/2}} \exp(-s^2/2)\, ds \right] dx. \tag{6}$$

The problem simplifies slightly if we employ $z = x(2DT)^{-1/2}$ as the variable of integration, with which, by (A8.11), our problem becomes the evaluation of

$$(2DT)^{1/2} C_0 A \int_{0}^{\infty} F(z)\, dz, \tag{7}$$

where

$$F(z) = 2 \left[1 - \frac{1}{\sqrt{(2\pi)}} \int_{-\infty}^{z} \exp(-s^2/2)\, ds \right]. \tag{8}$$

To evaluate $\int_0^\infty F(z)\, dz$, we employ the integration-by-parts formula (A8.13), replacing x by z, with the identification

$$g(z) \equiv F(z), \quad \frac{dh(z)}{dz} \equiv 1 \quad [\text{so that } h(z) = z]. \tag{9}$$

By the fundamental theorem of calculus (A8.8),

$$\frac{dF(z)}{dz} = \frac{-2}{\sqrt{(2\pi)}} \exp(-z^2/2),$$

so that integration by parts yields

$$\int_0^\infty F(z)\, dz = zF(z)\Big|_0^\infty + \frac{2}{\sqrt{(2\pi)}} \int_0^\infty z \exp(-z^2/2)\, dz. \tag{10}$$

The contribution of the first term on the right side of (10) is zero. That $zF(z)$ vanishes when $z = 0$ is obvious. It is also the case that

$$\lim_{z \to \infty} zF(z) = 0,$$

because the decay of $F(z)$ to zero is much faster than the growth of z itself. (The more knowledgeable in calculus can demonstrate this by applying l'Hospital's rule to $F(z)/z^{-1}$.) Because

$$\frac{2z}{\sqrt{(2\pi)}} \exp(-z^2/2) = \frac{d}{dz}\left[-\frac{2}{\sqrt{(2\pi)}} \exp(-z^2/2) \right],$$

we have, from (10),

$$\int_0^\infty F(z)\, dz = \left[-\frac{2}{\sqrt{(2\pi)}} \exp(-z^2/2) \right]_0^\infty = \frac{2}{\sqrt{(2\pi)}}. \tag{11}$$

The result (7.16) now follows by combining (7) and (11).

8

Developmental pattern formation and stability theory

A central problem in developmental biology is that of the genesis of spatial patterns. What makes leaves form their characteristic distribution, outlines, and vein structure? What mechanisms create the fivefold symmetry of a starfish, the stripes of a zebra, and the spots of a leopard? How do human embryos construct the correct number of correctly proportioned ribs and fingers? Questions of this type were mentioned by King Solomon in *Ecclesiastes* and have interested the wise until this day.

Background

There is an abundance of experimental results in developmental biology. To give just a little of their flavor, we shall consider a few experiments on hydra (Figure 8.1). This fresh-water creature is a fraction of a centimeter in length. It contains about 100,000 cells and can be crudely described as polar in structure, ranging from a head and tentacles to a basal disk. The hydra is a favorite system for experimental analysis because its structures regenerate rapidly (frequently within hours) after various amputations and transplantations.

Examples of findings on hydra are these (Wilby and Webster, 1970; Wolpert, Hicklin, and Hornbruch, 1971): (a) If the head is removed, and a small region from just below the head is grafted onto the body, then a head region regenerates. (b) If the grafted piece extends somewhat farther toward the basal disk (i.e., if the graft is less "headlike"), then two heads regenerate. (c) If the graft of (b) incorporates yet more head material, then again only one head regenerates. (d) If a head is amputated and transferred to the anterior portion of the gastric region, then a new head regenerates. (e) If a head is adjoined to the posterior gastric region and the original head is later removed, then no new anterior head regenerates. Later we shall discuss a theoretical explanation for these results, which are certainly puzzling at first glance.

It is generally accepted that biological patterns ultimately result from selective activation of genes. This broad principle does not illumine

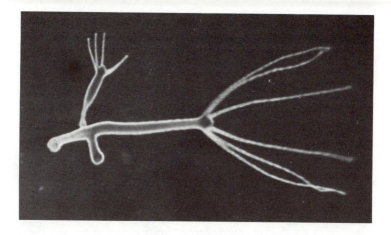

Figure 8.1. *Hydra littoralis,* with two buds. [From Wolpert, Hornbruch, and Clarke (1974).]

detailed experimental results such as those just described – we must explain why genes are activated in a specific spatiotemporal order and why this activation results in a specific structure. A theory is required, and in view of the range and complexity of the experimental results, we do not expect the theory to be elementary.

This chapter concentrates on one of several recent, somewhat related lines of theoretical work on pattern formation.[1] The initiation of this line is generally credited to the last paper written by A. M. Turing before his untimely death. As a young man, Turing was a key figure in the British success in cracking German communications codes during World War II. He is famous in scientific circles principally for what is now known as the Turing machine and related ideas in logic and the foundations of computer science. Becoming increasingly appreciated is his last paper. In this seminal work, Turing (1952) showed that under certain circumstances, chemicals that are simultaneously reacting and diffusing may automatically form a regular spatially inhomogeneous pattern, and he suggested that appropriate patterns of gene activation could result from such a chemical prepattern.

The 1970s saw the beginning of a vigorous development of Turing's ideas that continues to the present day. Salient contributions in this area have been made by A. Gierer and H. Meinhardt. Figure 8.2 shows an example of their line of research, in which special systems of postulates

---

1 The mathematical level is somewhat higher than that of the earlier material, particularly in the latter portion of this chapter.

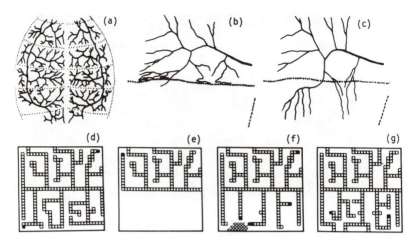

Figure 8.2. (a) A network of tracheae in an insect. (b) A segment with missing elements owing to disruption of the oxygen supply. (c) Regeneration after two weeks. [These results are due to Wigglesworth (1954).] (d)–(g) Regeneration of a net after half of it is removed, from computer simulation of a reaction-diffusion model. [From Meinhardt (1982), Figure 15.3, which should be consulted for further details.]

have been made concerning the nature of reactions so as to provide peaks of chemical concentration arranged in a form suitable for triggering or regenerating observed netlike patterns such as appear in leaf veins, capillary beds, or (in this case) insect trachael tubes.

Reference to the book of Meinhardt (1982) will provide the reader with a detailed account of the achievements of the Turing–Gierer–Meinhardt theory of pattern formation. At present, the theory can explain a wide variety of experiments, but in no particular case is there unequivocal identification of suitable chemicals in suitable concentrations. A major competing theory relies on mechanical forces to produce patterns and forms (Odell et al., 1981), but the mathematical settings of the two theories are similar in several respects. It seems safe to conclude, at the very least, that a researcher in developmental biology will profit by understanding the general lines of argument that we are about to present.

In what follows, we shall work through the analysis of a particular example of a Gierer–Meinhardt model. This will illustrate a number of the issues involved in various proposed mechanisms for pattern formation. Moreover, the mathematical analysis requires an extension of our earlier stability studies to situations in which the dependent variables depend on space as well as time. Such an extension provides a tool that is widely used in theoretical biology and in many other subjects.

We shall show that the model equations have a solution that corresponds to a uniform steady distribution of chemicals. We shall further demonstrate that under certain conditions this solution is unstable, in that spatially inhomogeneous perturbations to it will grow. The growth of these perturbations brings about a new nonuniform steady state that can provide a chemical prepattern for selectively switching on genes.

Formulation

It might be suggested that fertilization triggers the growth of a tiny organism that is already preformed in the egg. Many experiments have shown that this is not the case; instead, successive refinements and complexities acquired during a process of development eventually lead to the carefully proportioned three-dimensional adult. Thus, in establishing a theory of development, we must initially face the problem of generating a structure in a virtually featureless domain. As has been indicated, the Turing–Gierer–Meinhardt solution to this problem is the construction of pattern-generating chemical reactions. A key feature of such reactions is the presence of an autocatalytic activator, distinguished by a positive feedback on its own production. Autocatalysis alone generates explosive activity, not pattern. In addition, local activation must lead to the genesis and rapid spread of inhibitor – to damp the reaction in regions that are distant from foci of activity.

It turns out that the combination of short-range activation and long-range inhibition is sufficient to generate a variety of biologically interesting patterns. Given that this general approach is promising, to progress, we must devise equations whose solutions exhibit the properties required in various particular applications. Meinhardt (1982) gives many examples of equation systems with various pattern-formation properties. Success in deriving such equations comes from a combination of highly developed intuition plus many trials and errors with computer simulations. We shall not enter into this matter here, but rather take as given a basic equation of the Gierer–Meinhardt theory.

Let us assume for the present that chemical concentrations depend on a single spatial variable x and the time t. The concentration of an activator will be denoted by $a(x, t)$, and the concentration of an inhibitor will be denoted by $h(x, t)$. (In German, *Hemmstoffe* is the word for inhibitor, like "hemming-in stuff" in English.) Activator and inhibitor both diffuse, with fixed diffusivities D_a and D_h. The two chemicals are assumed to be secreted by cells whose densities (number/length) are, respectively, ρ_a and ρ_h.

The inhibitor is secreted at a rate $c_h a^2$ per cell, where c_h is a constant. This means that the inhibitor is produced only when there is catalysis by the activator. It turns out that the activator concentration must appear to some power (such as the a^2 used here) in order that the equations have the desired properties. Such "cooperative" dependence on catalyst concentration can be obtained from quasi-steady-state assumptions when catalyzing enzymes are activated only by the binding of more than one molecule of a (in this case, two such molecules).

It is postulated that there is activator production, at a constant rate c_a per cell, even at very low activator concentrations, but that activator also enhances its own production in the same fashion that it catalyzes the production of inhibitor. Further, it is assumed that the activation is opposed by inhibitor, so that the entire expression for the rate of activator production per cell is $c_a + ca^2/h$ for some constant c. Conventional expressions for inhibitor action would lead to a denominator of the form "constant $+ h$," instead of h itself, but the present form is sufficiently accurate if typical concentrations of h are sufficiently large.

Net productions of activator and inhibitor must potentially be in balance; otherwise there will be continual accumulation. This model assumes that activator and inhibitor are deactivated in the simplest possible way, proportional to their concentrations, at rates μ_a and μ_h.

With all the various assumptions, the model takes the form

◆◆
$$\frac{\partial a}{\partial t} = \rho_a\left(c_a + c\,\frac{a^2}{h}\right) - \mu_a a + D_a\,\frac{\partial^2 a}{\partial x^2}, \tag{1a}$$

$$\frac{\partial h}{\partial t} = \rho_h c_h a^2 - \mu_h h + D_h\,\frac{\partial^2 h}{\partial x^2}. \tag{1b}$$

For the time being we shall assume that the reaction takes place in a region that is in some sense "large," so that x can be taken to range between $-\infty$ and $+\infty$. We require the solutions to be finite as $x \to \pm\infty$.

We note that (1) is an example of what is called a **reaction-diffusion system** of equations. Such equations have recently been the object of intensive study. The analysis that we shall describe here is somewhat similar to (but in several ways more extensive than) material that appeared in a paper by Granero, Porati, and Zanacca (1977).

Dimensionless variables

We shall introduce dimensionless variables to make the algebra simpler (see Appendix 7). We choose as a time scale the reciprocal of μ_a, the decay constant for the activator. Thus, the dimensionless time τ will be defined by

$$\tau = t/\mu_a^{-1}, \quad \text{i.e., } \tau = \mu_a t. \tag{2}$$

As a length scale, we employ the magnitude of the distance through which the activator will diffuse in time μ_a^{-1}. From equation (7.20), this is $(D_a \mu_a^{-1})^{1/2}$, so that the dimensionless length s will be defined by

$$s = x/(D_a \mu_a^{-1})^{1/2}. \tag{3}$$

We define dimensionless activator and inhibitor concentrations A and H by

$$A = a/a_0, \quad H = h/h_0, \tag{4}$$

but for the moment we leave the constants a_0 and h_0 unspecified. On substituting the variables from (2), (3), and (4) into the governing equations (1a,b) and dividing by the coefficients of $\partial A/\partial \tau$ and $\partial H/\partial \tau$, we obtain

$$\frac{\partial A}{\partial \tau} = \frac{\rho_a c_a}{\mu_a a_0} + \frac{\rho_a c a_0}{\mu_a h_0} \frac{A^2}{H} - A + \frac{\partial^2 A}{\partial s^2}, \tag{5a}$$

$$\frac{\partial H}{\partial \tau} = \frac{\rho_h c_h a_0^2}{\mu_a h_0} A^2 - \frac{\mu_h}{\mu_a} H + \frac{D_h}{D_a} \frac{\partial^2 H}{\partial s^2}. \tag{5b}$$

We specify a_0 and h_o to achieve further simplification of the equations by demanding that

$$\frac{\rho_a c_a}{\mu_a a_0} = 1, \quad \frac{\rho_h c_h a_0^2}{\mu_a h_0} = \frac{\mu_h}{\mu_a} \tag{6}$$

so that

$$a_0 = \frac{\rho_a c_a}{\mu_a}, \quad h_0 = \frac{\rho_h \rho_a^2 c_a^2 c_h}{\mu_h \mu_a^2}. \tag{7}$$

The equations (5) now become

$$\frac{\partial A}{\partial \tau} = 1 + R \frac{A^2}{H} - A + \frac{\partial^2 A}{\partial s^2}, \tag{8a}$$

$$\frac{\partial H}{\partial \tau} = Q(A^2 - H) + P \frac{\partial^2 H}{\partial s^2}. \tag{8b}$$

The original equations (1) contained the nine parameters ρ_a, c_a, c, μ_a, D_a, ρ_h, c_h, μ_h, and D_h. The present equations contain just three dimensionless parameters:

◆◆
$$P = \frac{D_h}{D_a}, \quad Q = \frac{\mu_h}{\mu_a}, \quad R = \frac{c\mu_h}{c_a c_h \rho_h}. \tag{9}$$

Spatially homogeneous solutions

The underlying idea in the type of pattern-formation model under consideration is to find situations in which the uniform steady state is stable to spatially periodic perturbations of long and short wavelengths but unstable to perturbations of intermediate wavelengths. In such situations, by adjusting parameters, we can hope to find a spatially inhomogeneous pattern of an intermediate wavelength that matches the observed distance between repeating features such as hair follicles, feathers, or ribs.

A perturbation of very long wavelength varies slowly in space. Taking this situation to its limit, we put forward the requirement that for pattern-formation models there must be a stable steady solution in the spatially homogeneous case. To ensure this here, let us analyze the version of (8) that is appropriate when there is no spatial dependence $(\partial/\partial s = 0)$; namely,

$$\partial A/\partial \tau = f(A,H), \quad \partial H/\partial \tau = g(A,H), \tag{10}$$

where

$$f(A,H) = 1 + RA^2H^{-1} - A, \quad g(A,H) = Q(A^2 - H). \tag{11}$$

We now analyze (10) in exactly the same manner that we employed in Chapters 5 and 6.

The equations of (10) have a uniform steady solution

$$A = \bar{A}, \quad H = \bar{H}, \tag{12}$$

if

$$f(\bar{A}, \bar{H}) = 0, \quad g(\bar{A}, \bar{H}) = 0; \tag{13}$$

that is [Exercise 1(a)],

$$\bar{H} = \bar{A}^2, \quad \bar{A} = 1 + R. \tag{14}$$

Let us define perturbations $A'(\tau)$ and $H'(\tau)$ to the uniform steady state by

$$A(\tau) = \bar{A} + A'(\tau), \quad H(\tau) = \bar{H} + H'(\tau). \tag{15}$$

The linearized equations for $A'(\tau)$ and $H'(\tau)$ are [Exercise 1(a)]

$$\frac{dA'}{d\tau} = \frac{R-1}{R+1} A' - \frac{R}{(1+R)^2} H', \tag{16a}$$

$$\frac{dH'}{d\tau} = 2Q(1+R)A' - QH'. \tag{16b}$$

These equations are of the form

$$dx'/dt = ax' + by', \quad dy'/dt = cx' + dy', \tag{17}$$

a form that is extensively studied in Appendix 5. From Figure A5.9, the exponential solutions decay to zero if and only if $\gamma > 0$ and $\beta > 0$, where

$$\gamma = -Q\frac{R-1}{R+1} + \frac{R}{(1+R)^2} 2Q(1+R) = Q, \tag{18a}$$

$$\beta = -\left(\frac{R-1}{R+1} - Q\right). \tag{18b}$$

Because $Q > 0$, in this spatially homogeneous case stability is assured if and only if

$$\frac{R-1}{R+1} < Q. \tag{19}$$

We shall assume that (19) holds, so that the spatially homogeneous version of the activator–inhibitor interaction has a *unique constant solution* that is *stable to small spatially homogeneous perturbations*.

The spatially inhomogeneous case

We now turn our attention to the spatially inhomogeneous equations (8). The uniform steady (constant) solution (14) to (10) is still a solution to the present equation, for the new spatial derivative terms give zero when applied to a constant. Once again we define perturbations to the constant state, but now the perturbations will depend on space as well as time:

$$A(s,\tau) = \bar{A} + A'(s,\tau), \quad H(\tau,s) = \bar{H} + H'(s,\tau).$$

We again perform a stability analysis of the uniform state, but now with a much larger class of (nonuniform) perturbations.

The present version of the linearized equations (16) for A' and H' must now contain additional terms stemming from $\partial^2 A/\partial s^2$ and $P(\partial^2 H/\partial s^2)$. Because these terms are linear, it is readily seen that the counterpart of (16) in the spatially inhomogeneous case is

$$\frac{\partial A'}{\partial \tau} = \frac{R-1}{R+1} A' - \frac{R}{(1+R)^2} H' + \frac{\partial^2 A'}{\partial s^2}, \tag{20a}$$

$$\frac{\partial H'}{\partial \tau} = 2Q(1+R)A' - QH' + P\frac{\partial^2 H'}{\partial s^2}. \tag{20b}$$

The linearized perturbation equations (20) are more complicated than the corresponding equations that we have met earlier. They are partial differential equations, because the unknown functions A' and H' depend on two independent variables: the dimensionless spatial variable s as well as the dimensionless time τ.

We proceed by guessing a form of solution to (20).[2] We note that s occurs only in the form of second derivatives. Both sin and cos are proportional to their own second derivatives. Thus, either of these functions can be taken as the factor giving the s dependence of the solution, because this factor can be cancelled, as will be seen shortly. For definiteness, we use the cosine and assume

$$A'(s,\tau) = \hat{A}(\tau)\cos\frac{s}{l}, \quad H'(s,\tau) = \hat{H}(\tau)\cos\frac{s}{l}. \tag{21}$$

Here, l is a nonzero constant; $2\pi l$ is the period of the cosine. We shall see later that examination of the special solutions (21) supplies all the information we need to determine whether or not the uniform solution is stable.

To verify that (21) is a possible form of solution, let us substitute (21) into (20). We find after cancellation of the common factor $\cos(s/l)$ that the latter equations are indeed satisfied if

$$\frac{d\hat{A}}{d\tau} = \left(\frac{R-1}{R+1} - \frac{1}{l^2}\right)\hat{A} - \frac{R}{(1+R)^2}\hat{H}, \tag{22a}$$

$$\frac{d\hat{H}}{d\tau} = 2Q(1+R)\hat{A} - \left(Q + \frac{P}{l^2}\right)\hat{H}. \tag{22b}$$

To complete specification of the special solutions (21), we must determine the functions $\hat{A}$ and $\hat{H}$ from (22). Once again we are faced with the now-familiar problem of solving a system of two linear ordinary differential equations with constant coefficients. By Figure A5.9, the (exponential) solutions of (22) decay to zero if and only if

2 Our "guess" is in fact an easily verifiable reuse of an originally clever idea that has been passed from generation to generation. There is a close analogy to "guessing" the exponential form of solutions to ordinary differential equations with constant coefficients (Appendix 4).

$$-\left(\frac{R-1}{R+1}-\frac{1}{l^2}\right)\left(Q+\frac{P}{l^2}\right)+\frac{2QR}{R+1}>0, \tag{23a}$$

$$Q+\frac{P}{l^2}-\left(\frac{R-1}{R+1}-\frac{1}{l^2}\right)>0. \tag{23b}$$

Inequality (23b) definitely holds, because of assumption (19).

At this point we see that our special perturbations (21) will decay (stability) if (23a) holds and will grow otherwise (instability). To examine (23a) further, we multiply through by l^4 and rearrange. We obtain [Exercise 1(a)] as the equivalent of (23a) the *stability condition*

♦♦
$$Ul^4+\left(U-\frac{R-1}{R+1}\right)l^2+1>0, \tag{24}$$

where we have introduced the abbreviation

♦♦
$$U\equiv Q/P=\mu_h D_a/\mu_a D_h \tag{25}$$

for an important combination of parameters.

We observe that (24) is necessarily satisfied (stability) if

$$U>\frac{R-1}{R+1}. \tag{26}$$

But by (19) and (25), (26) will certainly hold if $P<1$. Because $P\equiv D_h/D_a$, we conclude that the uniform solution is stable, and no pattern can form, if $D_h<D_a$. Put in the opposite way, we see that spatial nonuniformity and the possibility of pattern formation require that the inhibitor diffuse faster than the activator. As we mentioned in our introductory remarks, it is quite a general phenomenon that *pattern formation requires short-range activation and long-range inhibition.*

Conditions for inhomogeneous instability

We now seek circumstances under which the uniform state is unstable. As a necessary condition, we assume that (26) is reversed:

$$\frac{R-1}{R+1}-U>0. \tag{27}$$

Reversal of the stability condition (24) gives

$$F(l^2)<0, \quad \text{where } F(\alpha)=U\alpha^2-\left(\frac{R-1}{R+1}-U\right)\alpha+1. \tag{28a,b}$$

(Here, α is just a dummy variable that we have used to define F.)

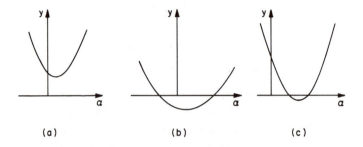

Figure 8.3. Some possible behaviors of the quadratic function $y = a\alpha^2 + b\alpha + c$, $a > 0$. (a) The quadratic has no roots: $y(\alpha) > 0$ for all α. (b) The quadratic has one positive root and one negative root: $y < 0$ for a range of negative and positive values of α. (c) The quadratic has two positive roots: $y < 0$ for a range of positive values of α.

The function $F(\alpha)$ just defined is quadratic, and it is positive for large values of α. Thus, its graph is a parabola that is concave upward. If $F(\alpha)$ is never zero, the graph is as shown in Figure 8.3(a), and $F(\alpha)$ is positive for all α. If $F(\alpha)$ is ever to be negative for positive $\alpha \equiv l^2$, then it is necessary that the equation $F(\alpha) = 0$ have two real roots, at least one of which is positive. See Figure 8.3(b,c). From (28), $F(\alpha) = 0$ when

$$2U\alpha = \frac{R-1}{R+1} - U \pm \left[\left(\frac{R-1}{R+1} - U\right)^2 - 4U\right]^{1/2}. \tag{29}$$

The roots of this equation are real when

$$\left(\frac{R-1}{R+1} - U\right)^2 > 4U. \tag{30}$$

When this is the case, (27) implies that both roots are positive, because $U > 0$. Because (27) holds, then on taking the square root of (30), we obtain

$$\frac{R-1}{R+1} > 2U^{1/2} + U. \tag{31}$$

Inequality (31) [which implies (27)] can be called an **instability condition**, because, as we have just seen, it guarantees that $F(\alpha) < 0$ for an interval of positive values of α – as is required by (28).

Equation (31) may be interpreted either as a condition on R or as a condition on U. Let us first examine the former possibility. If we assume that $2U^{1/2} + U < 1$, then [Exercise 1(a)] the instability condition (31) is equivalent to $R > R_c$; that is [from (9)],

$$\frac{c\mu_h}{c_a c_h \rho_h} > R_c, \tag{32}$$

where

$$R_c = \frac{1+(2U^{1/2}+U)}{1-(2U^{1/2}+U)}. \tag{33}$$

The key factor here is the coefficient c that governs the strength of the term in the governing equation (1a) (proportional to a^2/h) in which the activator accelerates its own production. For fixed U, then, sufficient autocatalysis will trigger instability.

In the examples that we shall consider later, there is strong autocatalysis, so that $R \gg 1$. In such a case, the instability condition (31) becomes, approximately,

$$U+2U^{1/2}<1. \tag{34}$$

We manipulate this inequality as follows:

$$U+2U^{1/2}+1<2, \quad (U^{1/2}+1)^2<2, \quad U^{1/2}<\sqrt{(2)}-1. \tag{35a,b,c}$$

Thus, for strong autocatalysis, the approximate instability condition (34) can be written, to good accuracy,

$$U^{1/2}<0.4, \quad \text{i.e.,} \quad \left(\frac{D_a/\mu_a}{D_h/\mu_h}\right)^{1/2}<0.4. \tag{36a,b}$$

Condition (36b) can be given an illuminating interpretation with the aid of the rule of thumb (7.20) for diffusion. Consider the numerator of (36b). Here the factor μ_a is the decay coefficient for activator in the governing equation (1a). If no other processes but decay influence the activator, the terms

$$da/dt = -\mu_a a \tag{37}$$

imply that $a \sim \exp(-\mu_a t)$, yielding the often-used observation that significant decay takes place in the time μ_a^{-1}. (Compare the discussion of time scale in Chapter 4.) From (7.20) it follows that an initial assemblage of activator molecules will diffuse a distance of magnitude $(D_a/\mu_a)^{1/2}$ before they disappear owing to their decay. Thus, we can usefully term $(D_a/\mu_a)^{1/2}$ the **activator range**. Similarly, $(D_h/\mu_h)^{1/2}$ is the **inhibitor range**. We conclude that *with strong autocatalysis, the condition for instability of the uniform state is that the inhibitor range be at least 2.5 times the activator range.*

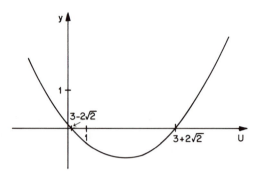

Figure 8.4. Graph of $y = U^2 - 6U + 1$.

We have interpreted the instability condition (31) alternatively as requiring sufficiently large R or sufficiently small U. Combining these interpretations, we conclude that, in general, instability requires a combination of sufficiently strong autocatalysis and sufficiently greater range for inhibition, as compared with activation.

The critical wavelength

For the remainder of this chapter we shall continue to restrict ourselves to the case of strong autocatalysis and assume that $R \gg 1$. This will simplify the algebra and will not alter the matters of principle that we wish to discuss.

We shall now examine the values of the wavelength parameter l of (21), corresponding to which, under various conditions, there are growing perturbations. When $R \gg 1$, we can approximate the instability condition (28) by

$$F(l^2) < 0, \quad \text{where } F(l^2) = Ul^4 - (1-U)l^2 + 1. \tag{38a,b}$$

The quadratic equation for l^2, $F(l^2) = 0$, has the two roots given by

$$l^2 = \frac{1 - U \pm [(1-U)^2 - 4U]^{1/2}}{2U}. \tag{39}$$

As we pointed out in Figure 8.3, there is a possibility that the concave-upward quadratic $F(\alpha)$ will be negative when $\alpha = l^2$ only if the roots given in (39) are real, which requires

$$(1-U)^2 - 4U > 0, \quad \text{i.e., } U^2 - 6U + 1 > 0. \tag{40}$$

A graph of $U^2 - 6U + 1$ is shown in Figure 8.4. We see that there are two

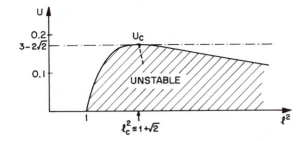

Figure 8.5. Values of (l^2, U) in the shaded region satisfy (38a). They therefore correspond to growing perturbations proportional to $\cos(s/l) \equiv \cos[x/l\sqrt{(D_a/\mu)}]$. The dashed line gives the locus of values $l_M(U)$ yielding the maximum growth rate, according to the approximate formula (90).

ranges of the nonnegative parameter U that permit real roots. The first range is

$$U > 3 + 2\sqrt{(2)},$$

but (38a) will not hold for such values of U, because each term in (38b) is positive. The other possibility is

$$0 < U < 3 - 2\sqrt{(2)}. \tag{41}$$

In this case the two roots of (39) are real and positive. The situation corresponds to Figure 8.3(c); inequality (38a) is satisfied (instability) for all values of l^2 such that

$$\frac{1 - U - [(1-U)^2 - 4U]^{1/2}}{2U} < l^2 < \frac{1 - U + [(1-U)^2 - 4U]^{1/2}}{2U}. \tag{42}$$

Note that although it has been derived in a different manner, (41) is equivalent to (35c), for $[\sqrt{(2)} - 1]^2 = 3 - 2\sqrt{(2)}$.

Let us consider the extremes of the range of U given by (41). When $U = 3 - 2\sqrt{(2)}$, the range of l^2 values in (42) collapses to a point. When $U \to 0$, on the other hand, we see [Exercise 1(b)] that the range in (42) is approximated by

$$1 < l^2 < U^{-1}, \quad 0 < U \ll 1. \tag{43}$$

We are thus led to the diagram of Figure 8.5 in our search for the values of l^2 that correspond to unstable disturbances proportional to $\cos(s/l)$.

We see from Figure 8.5 that if $U > 3 - 2\sqrt{(2)}$, then there are no growing disturbances. We define a critical value of U, U_c, by

$$U_c \equiv 3 - 2\sqrt{(2)} \approx 0.172. \tag{44}$$

Figure 8.5 shows that when U falls below U_c, the uniform state becomes unstable to disturbances proportional to $\cos(s/l)$, for a range of l.
When U just barely falls below U_c, only values of l^2 close to

$$l_c^2 = 1 + \sqrt{(2)}, \quad \text{i.e.,} \quad l_c \approx 1.55, \tag{45}$$

correspond to growing disturbances (Figure 8.5). Returning to our original spatial variable x, we note from (3) that

$$\cos\frac{s}{l} = \cos\frac{x}{l\sqrt{(D_a/\mu_a)}}, \tag{46}$$

so that these disturbances have a critical wavelength L_c given by

$$L_c = 2\pi l_c \sqrt{(D_a/\mu_a)} \approx 9.76\sqrt{(D_a/\mu_a)}, \tag{47a}$$

or, with sufficient accuracy,

◆◆ $$L_c \approx 10\sqrt{(D_a/\mu_a)}. \tag{47b}$$

We have reached the important conclusion that *when the instability condition $U < U_c$ is just barely satisfied, our linear theory predicts the exponential growth of a disturbance whose spatial period is the critical wavelength L_c given by (47).* We see that (with our assumption of strong autocatalysis) the critical wavelength is about an order of magnitude larger than the activator range.

Numerical analysis of the unstable case

What happens to the exponentially growing perturbations when they are no longer small? To answer this question we shall again follow a procedure that has been repeatedly exemplified in this book. A linearized stability analysis of the uniform state has been performed. The results of this analysis will now be employed to guide us in the choice of parameters that should be associated with particularly interesting solutions. These solutions will be investigated by means of a numerical analysis of the full nonlinear problem.

   In fact, we shall proceed in a slightly oblique fashion. Meinhardt and Gierer (1974) have already carried out a rather extensive numerical analysis of the type we are interested in. We shall be able to provide an ex post facto justification of their choice of parameters, using the linear theory. More important, from their computer study we shall learn the ultimate fate of the initially growing perturbations.

A numerical approximation to differential equations is effected by replacing derivatives by approximating difference quotients. (The problem is then suitable for a computer, which can easily perform arithmetic operations but generally cannot directly handle calculus problems.) A standard approach employs the approximations

$$\frac{\partial a(x,t)}{\partial t} \approx \frac{a(x,t+\Delta t) - a(x,t)}{\Delta t},$$

$$\frac{\partial a(x,t)}{\partial x} \approx \frac{a(x+\frac{1}{2}\Delta x,t) - a(x-\frac{1}{2}\Delta x,t)}{\Delta x}. \tag{48a}$$

From (48a),

$$\frac{\partial^2 a(x,t)}{\partial x^2} \approx \frac{\partial a(x+\frac{1}{2}\Delta x,t)/\partial x - \partial a(x-\frac{1}{2}\Delta x,t)/\partial x}{\Delta x}$$

$$\approx \frac{a(x+\Delta x,t) - 2a(x,t) + a(x-\Delta x,t)}{(\Delta x)^2}. \tag{48b}$$

Let us introduce the notation

$$a(i\Delta x,t) \equiv a_i(t), \quad h(i\Delta x,t) = h_i(t), \quad i = 0,1,2,\ldots. \tag{49}$$

If the derivatives in the original differential equations (1) are replaced by the approximations of (48), we then obtain

$$a_i(t+\Delta t) - a_i(t) = \rho_a[c_a' - c'(a_i^2/h_i)] - \mu_a' a_i + D_a'(a_{i+1} - 2a_i + a_{i-1}), \tag{50a}$$

$$h_i(t+\Delta t) - h_i(t) = \rho_h c_h' a_i^2 - \mu_h' h_i + D_h'(h_{i+1} - 2h_i + h_{i-1}), \tag{50b}$$

where

$$c_a' = c_a(\Delta t), \quad c' = c(\Delta t), \quad c_h' = c_h(\Delta t), \quad \mu_a' = \mu_a(\Delta t), \quad \mu_h' = \mu_h(\Delta t),$$

$$D_a' = D_a \Delta t/(\Delta x)^2, \quad D_h' = D_h \Delta t/(\Delta x)^2. \tag{51}$$

All functions on the right side of (50a) and (50b) are understood to be evaluated at t, so that these equations permit computation of $a_i(\Delta t)$ and $h_i(\Delta t)$ in terms of the initial conditions, then successively $a_i(2\Delta t)$ and $h_i(2\Delta t)$, $a_i(3\Delta t)$ and $h_i(3\Delta t)$, and so forth.

Equations of the type (50) are used by Meinhardt and Gierer in their extensive computer simulations. As we have seen, these equations can be obtained by approximating the reaction-diffusion equations (1). But the division of the x axis into compartments of width Δx can also be thought of as a model of the genuine cellular structure. Suppose that one makes the reasonable assumption that the flow rate of activator from one cell (compartment) to the next is some constant D'_a times the difference in activator concentration. In such a case, the new flow rate *into* cell i from its two neighbors is

$$D'_a(a_{i-1}-a_i)+D'_a(a_{i+1}-a_i)=D'_a(a_{i+1}-2a_i+a_{i-1}).$$

The preceding expression is identical with the last three terms in (50a). Instead of thinking of them as an approximation to more refined assumptions, we can regard the equations of (50) as a basic model depicting the reaction and intercellular flow of activator and inhibitor. If such a point of view is adopted, the differential equations (1) play the role of an approximation to the difference equations (50). Which of the two models is more fundamental depends on whether or not the cellular structure truly plays an essential role in the redistribution of the chemicals. (Chemical concentrations change continuously in time; so the discretization of time into intervals Δt must certainly be thought of as an approximation to the true state of affairs.)

The formation of a distributed set of activator and inhibitor peaks can be regarded as responsible for the initiation of a correspondingly distributed set of structures such as feathers, hairs, or spots. (Exposure to suitable levels of activator, or inhibitor, or a combination of both, can turn on appropriate genes.) Having found conditions under which small perturbations will result in the initial growth of such peaks, we wish to see whether or not a steady peak distribution will eventually result. Meinhardt and Gierer (1974) carried out the appropriate type of calculation. They wished to simulate the dispersal of peaks in a realistic two-dimensional situation. This requires the addition of terms proportional to $\partial^2 a/\partial y^2$ and $\partial^2 h/\partial y^2$ to (1a) and (1b) and addition of corresponding sets of terms to (50a) and (50b). The results of one of the Meinhardt and Gierer (1974) calculations are shown in Figure 8.6. With the parameters they employed, linear theory indeed predicts instability of the uniform state [Exercise 3(a)]. The numerical analysis shows that the initial exponential growth of small perturbations is followed by a slowing of growth and the eventual formation of a *steady* distribution of activator peaks.

It is to be expected that the critical wavelength should provide an estimate of the distance between activator peaks. In order to check our expectation, we must first adapt formula (47) for the critical wavelength L_c of the differential equations (1), to obtain an approximation to L_c for the *difference* equations (50). This is accomplished by substituting into (47) the relations (51) between the coefficients of the two equations, yielding

$$L_c=2\pi l_c\,\Delta x\sqrt{(D'_a/\mu'_a)},\quad\text{or }L_c\approx10\Delta x\sqrt{(D'_a/\mu'_a)}.\tag{52}$$

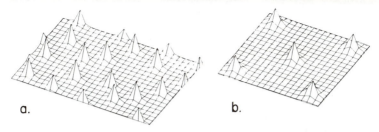

a.

b.

Figure 8.6. (a) Nonhomogeneous steady state exhibiting a fairly regular two-dimensional distribution of activator peaks. [From Meinhardt and Gierer (1974), Figure 7B.] To obtain this result, the two-dimensional versions of equations (50) were solved on a computer with parameter values $\rho_a = 1$, $c_a = 0.003$, $c = 0.01$, $\mu_a' = 0.015$, $D_a' = 0.003$, $\rho_h = 1$, $c_h' = 0.01$, $\mu_h' = 0.02$, $D_h' = 0.15$. Calculations began with random fluctuations in activator concentration and continued for many time steps until a steady pattern was formed. (b) Same as (a), except that growth is simulated. [From Meinhardt and Gierer (1974), Figure 7A.] The calculation begins with four elements. After every 400 iterations, new marginal elements are added, initially at the same chemical concentrations as their neighboring cells.

Note that in the discrete case the critical wavelength L_c is expressed as a multiple of the cell size Δx.

If the values of D_a' and μ_a' found in the legend for Figure 8.6 are substituted into (52), we obtain $L_c \approx 4.5\Delta x$. This indeed gives the correct magnitude of the observed interpeak spacing.

Linear theory predicted the growth of a sinusoidal perturbation, but it is apparent from Figure 8.6(a) that nonlinear effects distort the sinusoidal shape into a succession of fairly well defined peaks. Similar peaked shapes are seen more clearly in the one-dimensional simulation shown in Figure 8.7. In spite of the significant distortion that arises from nonlinear effects, the magnitude of the interpeak distance in this case is again adequately estimated by L_c [Exercise 3(c)]. This success is all the more noteworthy because Figure 8.7 was calculated under the supposition that there is a decrease in the source strength for activator and inhibitor from left to right. The observation that the source strength does not strongly influence interpeak spacing is consistent with the fact that ρ_a and ρ_h do not appear in formula (52) for the pattern wavelength according to linear theory. (Strictly speaking, our calculations do not apply for nonconstant sources, but replacing the actual source strengths by their average values often will yield qualitatively reasonable results.)

That the steady interpeak distance is not uniquely determined is shown by Figure 8.6(b), which was calculated under the same conditions as Figure 8.6(a) except growth was simulated by periodically adding marginal elements to an initially small (2 by 2) domain. Compared with

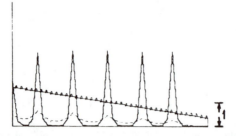

Figure 8.7. A row of activator (solid line) and inhibitor (dashed line) peaks calculated from (50). Linearly decreasing source distributions $\rho_a \equiv \rho_h$ are assumed (sawtooth line), with scale given at right. [From Gierer and Meinhardt (1972), Figure 1(i).] Parameters: $c_a = 0.0006$, $c = 0.05$, $\mu'_a = 0.01$, $D'_a = 0.001$, $c'_h = 0.025$, $\mu'_h = 0.01$, $D'_h = 0.04$.

Figure 8.6(a), the peaks are larger and more widely separated, and the distance between peaks is less well estimated by L_c.

Further remarks on the relationship between L_c and the interpeak distance will be made later.

Until now we have assumed that the spatial variable x extends from $-\infty$ to ∞. Actually, the interaction of activator and inhibitor will take place in a finite domain. Our assumption that the domain is infinite is equivalent to the assertion that the presence of boundaries will not appreciably affect the results. This assertion seems reasonable if the chemical wavelength of the predicted spatially nonuniform pattern is small compared with the size of the domain. In such a case we would anticipate that the boundaries would significantly distort the pattern only in their vicinity, but otherwise that the pattern would be quite similar to the predictions of the theory for the infinite region. Indeed, we have seen examples of computer calculations that show repeating patterns with a wavelength that is not far from that predicted by linear stability theory for an unbounded domain.

There are, however, important circumstances in which the length of the domain is comparable to the predicted instability wavelength. We turn to an examination of such a situation.

Patterns in finite intervals

Suppose that the spatial variable x is confined between $x=0$ and $x=L$ and that no chemical leaks through the boundaries. By expression (7.7) for the diffusive flux, the condition of *impermeable boundaries* is expressed by

$$\frac{\partial a}{\partial x} = \frac{\partial h}{\partial x} = 0 \quad \text{at } x=0 \text{ and } x=L. \tag{53}$$

With the introduction of the dimensionless variables defined in (3) and (4), (53) becomes

$$\frac{\partial A}{\partial s} = \frac{\partial H}{\partial s} = 0 \quad \text{at } s=0 \text{ and } s=\lambda, \tag{54}$$

where the dimensionless length parameter λ is given by

$$\lambda = L/(D_a \mu_a^{-1})^{1/2}. \tag{55}$$

We now review our previous analysis of the dimensionless equations (8), taking into account the new boundary conditions (54).

Because the derivative of a constant is zero, the uniform solutions (12) satisfy the boundary conditions and thus are permissible solutions to the modified problem. Moreover, the addition of boundary conditions does not require any modification of the linearized perturbation equations (20).

Again we seek solutions of the form

$$A'(s,\tau) = \hat{A}(\tau)\cos(s/l), \quad H'(s,\tau) = \hat{H}(\tau)\cos(s/l).$$

As previously, we limit ourselves to what we shall now term the *standard case* of large R, U slightly below U_c; that is,

$$R \gg 1, \quad 0 < U_c - U \ll 1. \tag{56}$$

Thus, the uniform state will be unstable when $l \approx l_c$, where l_c is still given by (45). Now there is an additional requirement, the boundary conditions (54). These require that

$$\sin(\lambda/l) = 0;$$

that is,

$$\lambda/l = k\pi, \quad k=1,2,3,\dots. \tag{57}$$

Using relationship (55) between λ and the interval length L, we find that the two requirements (57) and $l \approx l_c$ combine to give

$$L \approx k(D_a \mu_a^{-1})^{1/2}\pi l_c, \quad k=1,2,3,\dots. \tag{58a}$$

In terms of the critical wavelength L_c of (47), this condition can be written

$$L \approx k(L_c/2), \quad k=1,2,3,\dots. \tag{58b}$$

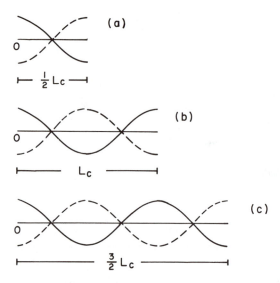

Figure 8.8. Demonstration that portions of the graph of $\cos(2\pi x/L_c)$ can "fit" into a region of length equal to an integer multiple of $L_c/2$ in such a way that the derivative with respect to x vanishes at the boundaries. In (a), the solid line is the graph of one-half of a single period of $\cos(2\pi x/L_c)$. The dashed line gives the graph of $-\cos(2\pi x/L_c)$. Correspondingly, the graphs of (b) and (c) give a single period and 1.5 periods, respectively, of $\cos(2\pi x/L_c)$.

In the form (58b), the condition we have derived is readily understood. Given our standard case (56), in an unbounded region a sinusoidal disturbance of wavelength approximately equal to L_c will grow. But for a region of finite length L, Figure 8.8. shows that the derivative of such disturbances vanishes on the boundary [as required by (54)] only if L is approximately an integer multiple, k, of $L_c/2$. The larger k is, the more oscillations are present in the interval $(0, L)$.

The initial-value problem

We now consider somewhat formally the initial-value problem for the case $k = 1$ of (57) and (58) (i.e., the case $L = \frac{1}{2}L_c$), wherein the interval is such that the only growing perturbation is the "half-wave" of Figure 8.8(a). In so doing, we will provide evidence for our earlier assertion that sinusoidal perturbations are sufficiently general to allow us to deal with quite broad initial conditions. Not surprisingly, we find that although an infinite series of sinusoidal functions is present initially, all of the terms in the series soon decay, except for the growing perturbation that we have already identified.

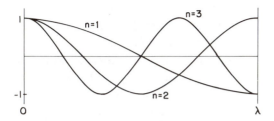

Figure 8.9. Graphs of $\cos(n\pi s/\lambda)$, $n=1,2,3$.

Let us now carefully set up a fully prescribed mathematical problem for the (dimensionless) perturbations $A'(s,\tau)$ and $H'(s,\tau)$. A typical reaction-diffusion problem requires for its complete specification (a) differential equations that describe the physical processes under consideration, (b) boundary conditions, and (c) initial conditions. As before, (20) provides the differential equations and (54) the boundary conditions. We also prescribe initial conditions at a time that we designate as $t=0$:

$$A'(s,0)=p(s), \quad H'(s,0)=q(s). \tag{59}$$

Here, p and q are given functions. We assume that there is no initial flow of material through the boundaries, so that p and q also satisfy the boundary conditions (54).

We attempt to solve the problem consisting of (20), (54), and (59) by considering solutions that are proportional to $\cos(s/l)$ for some constant l. (It may seem that we are repeating earlier analysis, but we are now proceeding more carefully and generally.) The boundary conditions (54) require that $\sin(\lambda/l)=0$, so that

$$\lambda/l=n\pi, \quad \text{i.e., } l=\lambda/n\pi.$$

Thus, we shall examine solutions of the form

$$A'(s,\tau)=\hat{A}_n(\tau)\cos\frac{n\pi s}{\lambda}, \quad H'(s,\tau)=\hat{H}_n(\tau)\cos\frac{n\pi s}{\lambda}, \tag{60}$$

where our notation now emphasizes that there are different solutions for each n. Because

$$\cos(-n\pi s/\lambda)=\cos(n\pi s/\lambda),$$

we do not need to consider negative values of n.

Graphs of $\cos(n\pi s/\lambda)$ for $n=1$, 2, and 3 are shown in Figure 8.9. The derivatives of all the chosen perturbations vanish at the boundaries, as required. It is seen that $n=1$ is the half-wave perturbation whose amplitude will grow, because the perturbation wavelength is the critical dimensionless wavelength πl_c [which comes from (57), with $l = l_c$, $k = 1$]. When $n = 2$, the perturbation wavelength is one-half the critical value, so that the amplitude of this perturbation is expected to decrease with time. Similarly, when $n = 3$, the wavelength is $\frac{1}{3}\pi l_c$, even shorter, so that this perturbation is also expected to decay. It is instructive to contrast Figure 8.8 with Figure 8.9. The former shows how perturbations of the critical wavelength can "fit" into different-sized regions; the latter indicates that a number of perturbation modes can fit into a region of given length – with decay expected for the "higher," more oscillatory, modes.

The linearized equations (20) and the boundary conditions (54) are satisfied by functions of the form (60), provided that the perturbation amplitude equations (22) hold for each $\hat{A}_n(\tau)$ and $\hat{H}_n(\tau)$, $n = 1, 2, 3, \ldots$, that is, provided that

$$\frac{d\hat{A}_n}{d\tau} = \left(\frac{R-1}{R+1} - \frac{n^2\pi^2}{\lambda^2} \right)\hat{A}_n - \frac{R}{(1+R)^2}\hat{H}_n, \tag{61a}$$

$$\frac{d\hat{H}_n}{d\tau} = 2Q(1+R)\hat{A}_n - \left(Q + \frac{Pn^2\pi^2}{\lambda^2} \right)\hat{H}_n. \tag{61b}$$

We know from the discussion in Appendix 4 that these linear equations with constant coefficients have solutions of the form

$$\hat{A}_n(\tau) = A_{n1}\exp(m_{n1}\tau) + A_{n2}\exp(m_{n2}\tau), \tag{62a}$$

$$\hat{H}_n(\tau) = C_{n1}A_{n1}\exp(m_{n1}\tau) + C_{n2}A_{n2}\exp(m_{n2}\tau),$$

$$n = 1, 2, 3, \ldots. \tag{62b}$$

Here, A_{n1} and A_{n2} are arbitrary constants, and the constants C_{n1} and C_{n2} are given by the following formula [Exercise 4(a)]:

$$C_{ni} = \frac{(1+R)^2}{R}\left(\frac{R-1}{R+1} - m_{ni} - \frac{n^2\pi^2}{\lambda^2} \right), \quad i = 1, 2. \tag{63}$$

The *growth rates* m_{n1} and m_{n2} are the two roots of the quadratic equation for m_n [Exercise 4(a)].

$$\left(\frac{R-1}{R+1} - \frac{n^2\pi^2}{\lambda^2} - m_n \right)\left(Q + \frac{Pn^2\pi^2}{\lambda^2} + m_n \right) - \frac{2QR}{1+R} = 0. \tag{64a}$$

We shall limit our discussion to situations in which the two roots m_n are real, in which case we denote the larger by m_{n1} and the smaller by m_{n2}:

$$m_{n1} \geqslant m_{n2}. \tag{64b}$$

Thus, the linear problem composed of the linear partial differential equations (20) and (54) has a whole family of solutions, a pair for each positive integer n. There is an additional special solution that is independent of s, corresponding to $n=0$:

$$A'(s,\tau) = A_{01} \exp(m_{01}\tau) + A_{02} \exp(m_{02}\tau), \tag{65a}$$

$$H'(s,\tau) = C_{01}A_{01} \exp(m_{01}\tau) + C_{02}A_{02} \exp(m_{02}\tau). \tag{65b}$$

Again, A_{01} and A_{02} are arbitrary, and C_{01}, C_{02}, m_{01} and m_{02} can be determined [Exercise 4(b)]. In fact, this solution is precisely the spatially uniform perturbation that we discussed at the beginning of this chapter. In particular, by requirement (19), we have ensured the decay of spatially homogeneous perturbations, so that the growth rates of (65) are negative:

$$m_{01} < 0, \quad m_{02} < 0. \tag{65c}$$

Because our present problem is linear, sums of solutions remain solutions. In an attempt to satisfy the initial conditions (59), we add together all the solutions that we have obtained in (60) and (62), together with (65) (thereby forming linear combinations of the type mentioned in Appendix 4):

$$A'(s,\tau) = \sum_{n=0}^{\infty} [A_{n1} \exp(m_{n1}\tau) + A_{n2} \exp(m_{n2}\tau)] \cos\frac{n\pi s}{\lambda}, \tag{66a}$$

$$H'(s,\tau) = \sum_{n=0}^{\infty} [C_{n1}A_{n1} \exp(m_{n1}\tau) + C_{n2}A_{n2} \exp(m_{n2}\tau)] \cos\frac{n\pi s}{\lambda}. \tag{66b}$$

The initial conditions (59) require that

$$\sum_{n=0}^{\infty} (A_{n1} + A_{n2}) \cos\frac{n\pi s}{\lambda} = p(s), \tag{67a}$$

$$\sum_{n=0}^{\infty} (C_{n1}A_{n1} + C_{n2}A_{n2}) \cos\frac{n\pi s}{\lambda} = q(s). \tag{67b}$$

It turns out (Exercise 5) that the requirements (67) can be formally met, provided that the coefficients A_{n1} and A_{n2} are the (unique) solutions of the pairs of equations

$$A_{n1} + A_{n2} = p_n, \tag{68a}$$

$$C_{n1}A_{n1} + C_{n2}A_{n2} = q_n. \tag{68b}$$

Here, p_n and q_n are given by formulas involving integrals of the given functions $p(s)$ and $q(s)$, $n=0,1,2,\ldots$.

Recall now that we are considering the standard slightly unstable situation (56) and that the length of the interval is such that only the half-wave perturbation of Figure 8.8(a) will grow (i.e., $\lambda = \pi l_c$ or $L = L_c/2$). Together with (65c), this means (Exercise 6) that

$$m_{n1} < 0, \quad m_{n2} < 0, \quad n \neq 1; \quad m_{12} < 0; \quad \text{but } m_{11} > 0. \tag{69}$$

Consequently, of the terms in the infinite series of (66), only a term proportional to $\cos(\pi s/\lambda) = \cos(\pi x/L)$ will grow as time goes on; all the others will decay. This, in turn, means that whatever the form of the initial perturbations of (59), the linear theory predicts that with the passage of time the deviation from uniformity will be more and more dominated by a term proportional to $\cos(\pi x/L)$.

To put into perspective the consequences of the finding we have just made, we note that for many years developmental biologists have suggested that patterns can be explained with the aid of a "gradient" of some key chemical (Child, 1941). (In this context, a gradient is said to be present if the chemical concentration changes monotonically with the spatial coordinate.) Yet, such an explanation seemed to demand a source of the chemical at one end of the interval and a sink at the other end. And it appeared nearly as difficult to explain the formation of the source and sink as it was to explain the patterns that were hypothesized to flow from the existence of the gradient. We can now see, however, that with the activator–inhibitor model, the gradient can be set up automatically.

Suppose, for definiteness, that a piece of tissue has length exactly equal to $\frac{1}{2}L_c$ and that U is slightly greater than U_c. Imagine that conditions slowly change so that U gradually drops below U_c. Let us make a mental "snapshot" of the situation at a certain instant $t = 0$, when U is just barely smaller than U_c. At that time there will be some distribution of activator and inhibitor, corresponding to which we can fix the coefficients A_{n1} and A_{n2} of the solution (66) from the equations (68). As we have seen, of all the terms in the infinite series of (66), only those proportional to $\exp(m_{11} t)$ will grow with time; all others will decay. The linear theory thus predicts that after a time the activator perturbation will be given by

$$A'(s, \tau) \approx A_{11} \exp(m_{11} \tau) \cos(\pi s/\lambda). \tag{70}$$

Choosing λ to fit the solution exactly into the length $L_c/2$, we find [Exercise 7(a)] that, in the original variables, the activator concentration itself satisfies

$$a(x, t) \approx \frac{\rho_a c_a}{\mu_a} \left[1 + R + A_{11} \exp(m_{11} \mu_a t) \cos(2\pi x/L_c) \right]. \tag{71}$$

Thus, depending on whether A_{11} is positive or negative, the activator perturbation will have the appearance of the solid or dashed curve in Figure 8.8(a). The sign of A_{11}, in turn, will be determined, generally speaking, by whether activator and inhibitor concentrations initially have a tendency to be higher on the left or on the right end of the domain [Exercise 7(b)]. In short, a gradient is "automatically" generated, with the location of its high point depending on conditions when U just barely drops below U_c.

Numerical analysis of the unstable case: finite interval

Our discussion has been based on the linear theory for small perturbations, but we know that the assumptions of this theory cease to be valid when a growing disturbance becomes sufficiently large. To obtain further understanding of instability in a relatively small domain, we turn to the computer.

Figure 8.10 is an elegant presentation of results by Meinhardt (1974) for the discrete system (50). [See the supplement to this chapter for remarks concerning the boundary conditions that must be used with (50).] The part of Figure 8.10 marked "a" shows homogeneous levels of activator and inhibitor that persist in time. These are the steady-state solutions. At "b," activator and inhibitor concentrations are decreased in a spatially homogeneous fashion. The perturbations are not particularly small, but they decay in accord with the predictions of linear theory. [Recall that inequality (19) ensures that the steady solutions are stable to homogeneous perturbations. According to Exercise 8(a), this inequality is satisfied for present parameter values.]

At "c" in Figure 8.10, activator and inhibitor concentrations are increased in a homogeneous fashion. Again, as expected, the calculations show a return to the steady state.

A small temporary pulse of activator concentration is administered at one end of the domain, at time "d." This nonhomogeneous perturbation triggers an instability, resulting in a new steady state with an activator (and inhibitor) peak with the same orientation as the original perturbation. The ultimate way that the activator concentration differs from uniformity is roughly like the portion of the cosine shown in Figure 8.8(a), but nonlinear effects have sharpened the peak, as they often do. Given that there is one peak, not two, reference to Figure 8.8 suggests that the length of the domain should be between $L_c/2$ and L_c, approximately. Indeed [Exercise 8(b)], with the parameters given in the legend for Figure 8.10,

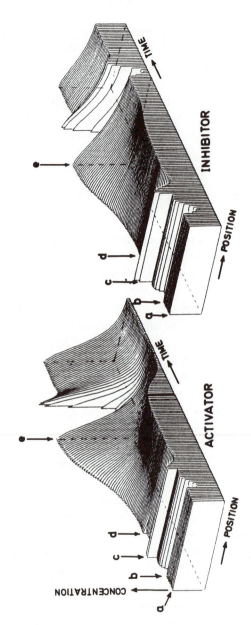

Figure 8.10. Development with time of activator and inhibitor distributions when the length of the domain is, approximately, between $\frac{1}{2}L_c$ and L_c. [From Meinhardt (1974), Figure 2.] We see the stability of the uniform state to uniform perturbations (a, b, c), the formation of a gradient owing to a small increase in activator concentration near the left boundary (d), and regeneration of the gradient after amputation of most of the activated area (e). Parameters: $c'_h = c' = 0.005$, $\rho_h = \rho_a = 1$, $c'_a = 0.0003$, $\mu'_a = 0.003$, $\mu'_h = 0.0075$, $D'_a = 0.008$, $D'_h = 0.45$.

$$L_c/2 = 8\Delta x, \tag{72}$$

and the length of the domain is either $15\Delta x$ or $11\Delta x$.

Regeneration of structures after amputation is a common phenomenon in lower organisms and embryos. Such a regeneration phenomenon is depicted in Figure 8.10. At time "e," the four cells containing the highest concentration of activator are "amputated." It is seen that a new activator peak is soon regenerated. In view of linear stability theory, this result should not come as a surprise. The conditions for instability still hold; so the remaining trend to higher activation concentration toward the left is amplified into a new steady state.

Let us now turn our attention to Figure 8.11(a), which depicts how a gradient can suddenly be generated in a growing region. This is precisely what we would expect from the stability results of Figure 8.5, which show that $l = 1$ is the smallest value that will permit instability. Employing (46), we see that $\pi\sqrt{(D_a/\mu_a)}$ is the minimum length of a domain permitting the observed half-wavelength instability of Figure 8.8(a). If the domain is of more than about twice this length, we anticipate that the activator concentration can take either of the forms of Figures 8.8(b) or 8.8(c), depending on initial conditions. That such is indeed the case is demonstrated in Figure 8.11(b) and (c). In a borderline case, when $L \approx 2\pi\sqrt{(D_a/\mu_a)}$, initial conditions or some other external perturbation should determine which of the patterns in Figure 8.11 will be assumed by the activator concentration. Meinhardt (1982, p. 18) suggests that this may be the reason why sometimes polarized light can induce a double rhizoid in *Fucus* algae (Figure 8.12).

We have now seen that a number of predictions that are strictly valid only for very small disturbances to the steady state in fact remain approximately true when disturbances become large. By contrast, a deficiency of linear theory is its prediction that unstable perturbations continue to increase exponentially in time. Fortunately, it appears that nonlinear effects often act to limit perturbation growth, resulting in a new and possibly useful spatially inhomogeneous steady state.

The wavelength for maximum growth rate

We wish now to attain a better understanding of the distances between activator and inhibitor peaks that appear in the computer simulations. To that end, in this somewhat technical section we derive the approximate formula (90b) for the dimensionless wavelength l_M that is associated with the perturbation that grows fastest at a given value of the parameter U. The implications of (90b) are discussed in the next section.

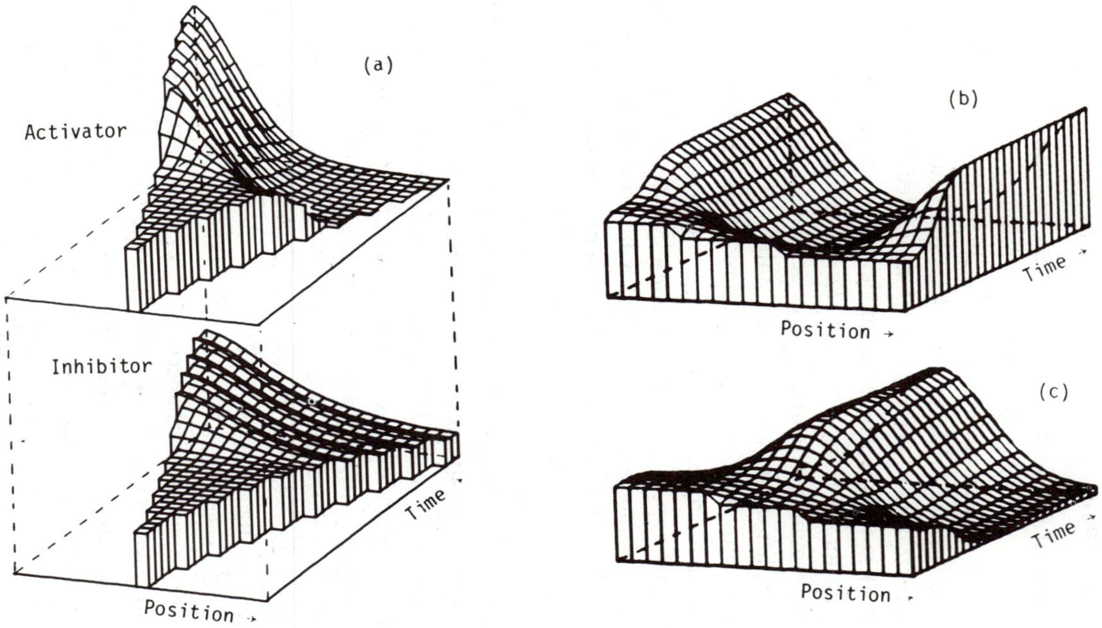

Figure 8.11. (a) Growth "releases" a gradient in the concentrations of activator and inhibitor. (b) In a larger domain, there appears an activator concentration that is similar to the growing perturbation represented by the solid line in Figure 8.8(b). (c) In the same size domain as in (b), a slight difference in initial conditions can lead to the dominance of the perturbation that is represented by the dashed line in Figure 8.8(b). [From Meinhardt (1982), Figure 4.1.]

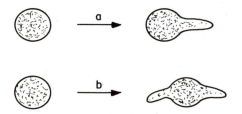

Figure 8.12. (a) Schematic representation of the normal outgrowth of a rhizoid from a *Fucus* egg. (b) A double rhizoid, as can sometimes be induced when polarized light is used to illumine the egg. If a high activator concentration triggers rhizoid formation, the essence of the situation may be explained by the alternatives of Figure 8.11(a) and 8.11(b).

We have seen that under certain circumstances the linear theory can predict the magnitude of the distance between activator peaks in a large domain, and also the size of a domain that will permit the spontaneous emergence of an activator gradient. These two predictions are, respectively, L_c and $L_c/2$ – under circumstances such as (56) that are just barely unstable according to the linear theory. What of cases for which the instability is more marked? Aside from its intrinsic interest, this question should be answered because for the three examples we have examined in detail, U is considerably less than its critical value (for large R) of $U_c \approx 0.17$. For example, for the simulations of Figure 8.10,

$$R = 25 \gg 1, \quad U = 0.044. \tag{73}$$

As is depicted in Figure 8.5, when U drops farther and farther below U_c, the values of l corresponding to a growing disturbance form a rapidly broadening band. What length scale characterizes the activator–inhibitor distribution in such cases? For a large domain, we hypothesize that the distance between activator peaks will be given approximately by the value of l that maximizes the largest of the growth rates m_{ni} in the assumed form of disturbance (66). We denote this value of l by l_M, and we turn now to the computation of l_M. In carrying out this computation, we shall forgo generality and accuracy for simplicity, by continuing to restrict ourselves (but not so strictly as before) to the case $U \approx U_c$.

From the present standpoint, our previous discussion can be regarded as showing that when U is slightly below U_c, then l_c is the value of l that corresponds to the maximum growth rate. For, in this situation, if $l \neq l_c$, then all the growth rates m_n of (64a) are negative, whereas m_{11} is positive if $l \approx l_c$. We now wish to make a more accurate determination of the value l_M of l that maximizes m_{11}. To this end we shall define a small parameter ϵ by

$$\epsilon \equiv U - U_c. \tag{74}$$

Regarding all parameters but U as fixed, we can think of l_M as a function of U or of ϵ. We select $l_M(\epsilon)$ in order to maximize $m_{11}(\epsilon)$. As we have discussed, when ϵ tends to zero, then $l_M(\epsilon)$ tends to l_c; that is,

$$l_M(0) = l_c. \tag{75}$$

Note that when U is slightly below U_c (instability),

$$\epsilon < 0, \quad |\epsilon| \ll 1. \tag{76}$$

The quantity m_{11} that we wish to maximize is the larger root of equation (64a) when $n = 1$. Thus, once again, assuming $R \gg 1$ for simplicity, we see that m_{11} and m_{12} satisfy [Exercise 9(a)]

$$m_1^2 + [Q + Pl^{-2} - 1 + l^{-2}]m_1 + Pl^{-4}[Ul^4 + l^2(U-1) + 1] = 0. \tag{77}$$

When (76) holds, the smaller root of (77), m_{12}, is a negative number, and the larger root, m_{11}, is a very small positive number [Exercise 9(b)]. To a first approximation, then, we can neglect the term m_{11}^2 as being negligible compared with the term proportional to m_{11} and write

$$m_{11} = -g(U, l^2)/h(U, l^2), \tag{78}$$

where

$$g(u, l^2) \equiv Ul^4 + l^2(U-1) + 1, \tag{79}$$

$$h(U, l^2) = P^{-1}[(UP-1)l^4 + (P+1)l^2]. \tag{80}$$

We do not explicitly indicate the dependence on P, for we regard this parameter as fixed. The parameter Q has been replaced by UP, according to the definition of U in (25).

To find the value l_M of l that maximizes m_{11}, we equate to zero the derivative of m_{11} with respect to l^2 [Exercise 9(c)]. This yields

$$h(U, l^2)g_2(U, l^2) - g(U, l^2)h_2(U, l^2) = 0, \tag{81}$$

where the subscript 2 indicates that a partial derivative is being taken with respect to the second argument of the function. For example,

$$g_2(U, l^2) = \partial g(U, l^2)/\partial l^2. \tag{82}$$

If we substitute $U = U_c + \epsilon$ [from (74)], then (81) provides the desired equation for l_M^2 as a function of ϵ. We know from (75) that the solution will be close to l_c^2 when ϵ is small. We thus write

$$l_M^2 = l_c^2 + \delta, \tag{83}$$

and we anticipate that δ will be small.

Incorporation of (74) and (83) into (81) yields

$$0 = h(U_c + \epsilon, l_c^2 + \delta) g_2(U_c + \epsilon, l_c^2 + \delta) - g(U_c + \epsilon, l_c^2 + \delta) h_2(U_c + \epsilon, l_c^2 + \delta). \tag{84}$$

Our task is to approximate the solution of this equation for small ϵ and δ. In carrying this out, we employ the Taylor formula in two variables (A2.8) to approximate the terms in (84). Denoting higher-order terms by (...), we obtain (Exercise 10)

$$0 = (h^c + \epsilon h_1{}^c + \delta h_2^c)(g_2^c + \epsilon g_{21}^c + \delta g_{22}^c + \dots)$$
$$- (g^c + \epsilon g_1^c + \delta g_2^c + \dots)(h_2^c + \epsilon h_{21}^c + \delta h_{22}^c + \dots), \tag{85}$$

where a superscript c on g or h implies that the function is evaluated at (U_c, l_c^2).

We have extended the notation of (82), writing, for example,

$$\frac{\partial^2 g(U, l^2)}{\partial l^2 \partial U}\bigg|_{U = U_c, l^2 = l_c^2} \equiv g_{21}^c.$$

At this point we pause to reexamine equation (78) for the largest growth rate m_{11}. This is necessary for further progress in our present investigation, but it also will contribute to our general understanding of stability theory.

In all cases that we consider, $Q \equiv UP \geqslant 1$ [in line with requirement (19) that ensures stability of the uniform state to spatially homogeneous perturbations]. Thus, the denominator $h(U, l^2)$ of (78) is positive, and the sign of m_{11} is determined by the numerator, $-g(U, l^2)$. The curve along which $m_{11} = 0$ separates the region in the (U, l^2) plane in which m_{11} is positive from the region in which m_{11} is negative. Thus, *the curve bounding the unstable region in Figure 8.5 is the locus of points where $m_{11} = 0$.* The equation for this curve was given earlier in (39); in present notation, the equation is $g(U, l^2) = 0$. It follows (Exercise 10) that

$$g^c = 0, \quad g_2^c = 0. \tag{86}$$

Employing the crucial relations (86), and neglecting higher-order terms proportional to ϵ^2, $\epsilon\delta$, and δ^2, we find that (85) reduces to

$$h^c g_{21}^c \epsilon + h^c g_{22}^c \delta - g_1^c h_2^c \epsilon = 0. \tag{87}$$

This provides the desired formula for δ:

$$\delta = \frac{g_1^c h_2^c - h^c g_{21}^c}{h^c g_{22}^c} \epsilon. \tag{88}$$

Substituting for g and h from (79) and (80) and carrying out the required differentiations and simplifications, we find [using (44) and (45)] that

$$\delta = \frac{l_c^2}{2U_c} \frac{P(U_c-1)-2}{P(U_c l_c^2+1)-l_c^2+1} \epsilon \approx -10 \frac{0.4P+1}{P-1} \epsilon. \tag{89}$$

In the various examples we have considered, $P \gg 1$. In this case we arrive at a final, very simple approximate formula

$$\delta = -4\epsilon, \quad \text{i.e.,} \quad l_M^2 = l_c^2 + 4(U_c - U). \tag{90a,b}$$

The resulting locus of points yielding maximum growth rate is given by the dashed line in Figure 8.5.

Further examination of pattern size

Let us now reexamine the question of what wavelength we expect to observe in an unstable situation, employing the result just obtained, but also other considerations. If U is only slightly smaller than U_c, then, as we have seen, the predictions of linear theory are quite decisive. A wavelength near L_c should be observed, because only disturbances of approximately this wavelength can grow; all other disturbances decay. If U drops somewhat below U_c, then we have suggested that the observed wavelength should be approximately that associated with the *maximum growth rate*. In accord with the approximation (90), as depicted in Figure 8.5, the simulation results depict a dominant wavelength that is similar to but somewhat larger than L_c, even though U is generally markedly below U_c, so that a very broad band of wavelengths has the potential for growth.

The dominance of the disturbance of maximum growth rate can be formally predicted from linear theory. To see this, let m_M denote the maximum growth rate; m will denote any other growth rate ($m < m_M$). Then the ratio of the corresponding disturbance amplitudes is proportional to $e^{mt}/e^{m_M t}$, which approaches zero as $t \to \infty$. This argument cannot be taken too seriously, however, because once disturbances reach an appreciable size, the assumption that perturbations are small ceases to be valid. Hitherto neglected nonlinear terms become important, so that some sort of *nonlinear stability theory* must be constructed. Such a theory has been developed during the last 25 years, and a vigorous research effort is continuing. We shall now employ several basic results of the nonlinear theory to shed light on the phenomena under consideration.

Suppose that the uniform state is initially perturbed by a purely sinusoidal disturbance or *mode* of wavelength L_c, and that $U < U_c$. According to linear theory, the disturbance amplitude continues to increase exponentially. However, the calculations on our problem by Granero and associates (1977) imply that nonlinear terms continually act to decrease the growth of the sinusoidal mode in question. Eventually a steady state is reached, consisting of the initial mode [proportional to $\cos(2\pi x/L_c)$] and its **harmonics** [proportional to $\cos(2\pi nx/L_c)$, $n = 2, 3, 4, \ldots$]. In general, there are also changes in the average activator and inhibitor concentrations. This illustrates *three major effects of nonlinearity: self-damping of a growing mode, generation of harmonics, and alteration of the spatial average.*

Now suppose, as is generally the case, that the initial perturbation consists of a number of sinusoidal modes, many of which grow according to the linear theory. *Another major effect of nonlinearity, intermode suppression,* now comes into play: The growth of one mode tends to damp the growth of other modes. It is as if a number of factors were competing for finite resources, and the success of one factor is necessarily compensated for by a depression of its competitors. Therein lies the true advantage of the fastest-growing mode. If all other considerations are equal, its growth depresses its competitors, particularly those that are growing relatively slowly. There is thus a nonlinear focusing of disturbance amplitude, the "strong" (relatively rapidly growing) modes gaining at the expense of the "weak" (relatively slowly growing) modes.

Aside from its growth rate m, each mode is associated with another important factor – its initial amplitude. The latter factor is not of fundamental importance according to linear theory, because a faster exponential growth will always outweigh the handicap of an initially smaller amplitude. When nonlinear effects are taken into account, the operation of intermode suppression implies that a mode of sufficiently dominant initial amplitude may extinguish modes with larger growth rates. The role of initial conditions can thus be decisive – which implies that the final state may not be unique. We have seen an example of such non-uniqueness in comparing Figures 8.6(a) and 8.6(b), two final steady solutions to the same set of equations. In this case, it was not initial conditions but rather differences in the size of the domain and in boundary conditions that distinguished the two problems. The general idea is the same, however. Intermode suppression can convert a temporary advantage into a permanent one.

We would expect, correctly as it turns out, that there is a limit to what intermode suppression can accomplish. Modes with growth rates that are too small compared with those of competitive modes will die out, no matter how large they are initially. This disadvantage of modes with relatively small growth rates is greater if we admit only "reasonable" initial conditions, with no enormous disparities in initial amplitudes among modes. This last line of reasoning lends further support to the assertion that the final steady state should normally consist of modes whose wavelengths are not too far from the wavelength that is associated with the maximum growth rate.

A supplement to this chapter contains a mathematical caricature of nonlinear modal interactions that clarifies some of the general remarks we have made. A connection is made between the struggle for survival between modes and the analogous struggle that goes on in evolution – at both the species and molecular levels. For those who are curious about nonlinear stability theory, a relatively elementary introduction can be found in a review article by Segel (1966). Coullet and Spiegel (1983) provide a good starting point for inquiry into recent developments.

Modeling experiments in hydra

The use of stability calculations coupled with computer simulations is a main theme of this book, and therefore this approach to the Gierer–Meinhardt theory has been emphasized. We now show briefly that once the basic activator–inhibitor pattern-forming mechanism is established, we can and should use extensive further computer simulations to come to grips with the wealth of available experimental results. To that end, we shall discuss the hydra experiments that were mentioned at the beginning of this chapter. For more information, see Meinhardt (1982, Chapter 6) and MacWilliams (1982).

Figure 8.13(a) is a useful schematic representation of the hydra, with its tentacles, head (H), a gastric region that is divided for convenience into four parts, a budding area (B), peduncle (P), and basal disc (D) (compare Figure 8.1). The two basic assumptions made by Gierer and Meinhardt (1972) in their analysis are as follows:

(a) There is an initially nonuniform distribution of activator and sources, resulting in secretion that is relatively high in the head, diminishing distally (toward the basal disc) throughout the gastric region.

(b) High activator concentration triggers head formation.

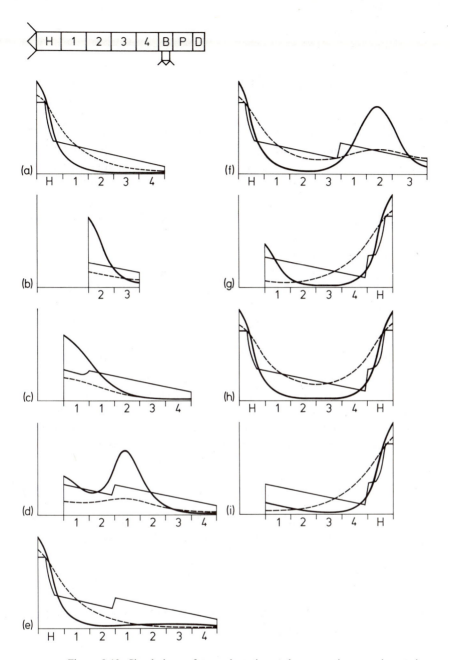

Figure 8.13. Simulations of transplantation and regeneration experiments in hydra. [From Gierer and Meinhardt (1972), Figure 2.] (a) Top: Schematic representation of hydra. Bottom: Distribution of source strengths $\rho_h = \rho_a$ (solid line), with full scale equal to 3.2 units, and resulting activator (heavy line) and inhibitor (dashed line) distributions. (b) Short gastric section regenerates a head. (c) Graft 1/1234 regenerates head. (d) Two heads regenerate from graft 12/1234. (e) Addition of H to transplant 12 inhibits second head. (f) Two heads regenerate

How can the gradient in source strength arise? There is no compelling need to answer this natural question until we see whether or not the assumption that a certain source gradient exists leads to a theory of strong explanatory power. If not, there is no need to proceed further. If so, at least we have pushed the frontier of our ignorance one step backward. Moreover, it is quite natural to start an explanation of development by postulating an existing nonuniformity that the initial stages of a new organism inherit from the parent. Here it is not unreasonable to anticipate that during its formation the bud inherits a graded distribution of secretory cells from its parent hydra. Moreover, the inherited gradient might be only a slight one that is magnified toward the head end by a separate activator–inhibitor system.

A graph of the assumed distribution of sources is shown in Figure 8.13(a). The legend for Figure 8.13 gives parameter values that were used in (50) to generate all the results that will now be described, results that are in conformity with amputation and grafting experiments of Wolpert and associates (1971) and Wilby and Webster (1970).

Figure 8.13(b) shows that if a relatively short piece of the gastric area is excised, an activator peak is developed. This result is expected from the stability theory in view of the parameter values that have been selected [Exercise 3(c)]. Calculations for 800 time intervals Δt were required to generate an activator peak. The total time $800\Delta t$ is assumed to correspond to the observed time of about four hours to determine a new head. Computer time is thus linked to real time by

$$\Delta t = 20 \text{ sec.} \tag{91}$$

In the calculations, the region H1234 is broken up into 40 units of length Δx. This region is typically about 4 mm long, giving the correspondence

$$\Delta x = 10^{-2} \text{ cm.} \tag{92}$$

Equations (91) and (92), coupled with relation (51) between D'_h and D_h, yield for the inhibitor diffusivity

$$D_h = 2.5 \times 10^{-6} \text{ cm}^2\text{-sec}^{-1}.$$

Caption to Figure 8.13 *(cont.)*
from still larger graft. (g) Successful transplantation of head from region 1 to region 4. (h) Successful transplantation of second head to region 4. (i) Ten hours after (h), original head is removed; no new head regenerates. Parameters: $c_a = 7.5 \times 10^{-4}$, $c = 0.05$, $\mu'_a = 0.0035$, $D'_a = 0.03$, $c'_h = 0.025$, $\mu'_h = 0.0045$, $D'_h = 0.45$.

This is a reasonable value; indeed, the value of D'_h was selected by starting with this value of D_h and working backward.

We now report the results of Gierer and Meinhardt (1972) that mimic the experimental findings (a)–(e) listed at the beginning of this chapter. First, if a proximal gastric region 1 is grafted onto the entire gastric region 1234, a new head is regenerated at the proximal end of the tissue [Figure 8.13(c)]. A more extensive graft 12/1234 results in *two* activator peaks and therefore, by our assumptions, two heads [Figure 8.13(d)]. The appearance of two peaks is, of course, to be expected in a sufficiently long piece of tissue.

An even more extensive graft, joining H12 to the gastric region 1234, does not yield a second head [Figure 8.13(e)]. The reason is that the inhibitor secretion from region H is able to counteract the "natural" tendency to form an internal activator peak that was present in the roughly homogeneous material represented in Figure 8.13(d). Figure 8.13(f) shows that an even more extensive graft, H123/123, once again results in the appearance of two heads, presumably because the high activator secretion in the head is now too far away to inhibit the natural formation of a second activator peak.

Two heads develop after the head is amputated and grafted to the anterior portion (region 4) of the gastric region [Figure 8.13(g)]. Suppose, however, that a new head is grafted to region 4 [Figure 8.13(h)], and the original head is removed only after 10 hours [Figure 8.13(i)]. Then the original head is not regenerated, because the posterior head has had sufficient time to propagate inhibitor.

MacWilliams (1982) performed an intensive numerical study of a version of equation (1) in which activator saturation was introduced by replacing a^2/h in (1a) by $a^2(h + \text{constant} \cdot a^2)$ and a^2 in (1b) by $a^2/(\text{constant} + a^2)$. The nine parameters of the model were estimated from 11 experiments; thereafter, the results of six further experiments were predicted.

Rapid progress in understanding hydra morphogenesis can be anticipated now that inducers and inhibitors have been purified (Schaller and Bodenmüller, 1982). Based on these experimental results, Kemner (1983) proposed a new model in which the Gierer–Meinhardt positive feedback of the activator on its own production (autocatalysis) is replaced by a negative feedback of h on *its* production (autoinhibition).

Two approaches to vein patterns in plants

The Turing–Gierer–Meinhardt theory of pattern formation has as one of its principal assumptions that suitable concentrations of certain key

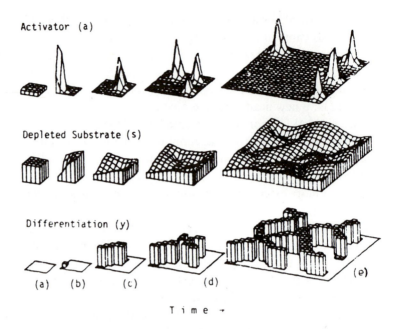

Activator (a)

Depleted Substrate (s)

Differentiation (y)

(a) (b) (c) (d) (e)

T i m e →

Figure 8.14. Dichotomously branching leaf pattern (found in the ginkgo tree, for example) generated by equations (93) in growing domain. [From Meinhardt (1982), Figure 15.6.]

chemicals (morphogens) will trigger developmental events. By now, there is a body of theory that shows how appropriate spatial patterns of morphogens can be made to emerge from the interplay of chemical reaction and diffusion. We have studied one example in considerable detail. Many other interesting examples are discussed by Meinhardt (1982), such as that of Figure 8.14. The goal here was to generate a netlike pattern of the type found, for example, in leaf veins, capillary beds, and bronchial tubes. Differentiation into net cells was identified with the irreversible switching of a chemical y from a low to a high steady state. Activator a generated the production of y, which in turn stimulated the production of a. In this model, inhibition arises from the depletion of a substance s that is necessary for the production of a. The particular equations that were used to generate the results of Figure 8.14 are

$$\frac{\partial a}{\partial t} = 0.008a^2 s - 0.04a + 0.0065\left(\frac{\partial^2 a}{\partial x^2} + \frac{\partial^2 a}{\partial y^2}\right), \tag{93a}$$

$$\frac{\partial s}{\partial t} = 0.05 - 0.008a^2 s - 0.25ys + 0.18\left(\frac{\partial^2 s}{\partial x^2} + \frac{\partial^2 s}{\partial y^2}\right), \tag{93b}$$

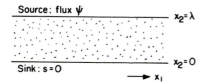

Figure 8.15. Representation of a fragment of plant tissue exposed to a flux of hormone.

$$\frac{\partial y}{\partial t} = 0.00032a - 0.1y + \frac{y^2}{1+10y^2}.$$ (93c)

(By now the reader can recognize the additional assumptions of these equations that we have not mentioned explicitly: diffusion, spontaneous decay, saturating autocatalytic production.) Also see Figure 8.2, showing the results of a simulation of equations involving an inhibitor as well as the three substances of (93). Here, a more complex pattern results when there is a separation of two required inhibitory actions: localization of activation at the vein tip and determination of the direction of tip migration.

In contrast with theories for which the *level* of a morphogen is decisive, Sachs (1969) suggested that for the differentiation of leaf veins it is the *rate of flow* that determines differentiation. Sachs's hypothesis is based on a number of experiments concerned with vein induction by the plant hormone auxin (indoleacetic acid). Mitchison (1980, 1981) has constructed and tested a mathematical model based on the Sachs hypothesis. We shall outline the first elements of this model here. An alternative approach to pattern formation will thereby be presented, but the fundamental concept of instability plays a key role here too.

Following the approach of Mitchison (1980), let us imagine a long piece of plant tissue, bounded by lines that we denote by $x_2 = 0$ and $x_2 = \lambda$ (Figure 8.15). Suppose that there is a source of hormone (concentration s) at $x_2 = \lambda$ and that the hormone diffuses across the tissue to a sink (perhaps a large vein) at $x_2 = 0$. Let D_i denote the diffusivity in the x_i direction, $i = 1, 2$. We shall assume that the diffusivity in a given direction will increase if the flux magnitude in that direction is larger. The question is, Under what conditions, if any, will such an assumption lead to the automatic generation of preferred channels of flow, that is, of smaller veins that will feed hormone (among other things) to the larger vein?

Let ϕ_i denote the flux in the direction of *decreasing* x_i. (It is convenient to modify the standard definition, wherein flux in the direction of increasing x_i is regarded as positive.) By (7.7),

$$\phi_i = D_i(\partial s/\partial x_i). \tag{94}$$

The two-dimensional version of diffusion equation (7.8) becomes

$$\frac{\partial s}{\partial t} = \frac{\partial \phi_1}{\partial x_1} + \frac{\partial \phi_2}{\partial x_2}. \tag{95}$$

Modification of the diffusivities is assumed to be governed by

$$\frac{\partial D_i}{\partial t} = f(|\phi_i|, D_i), \quad i = 1, 2. \tag{96}$$

To specify our assumptions on the nature of f, let us introduce a notation in which f_i denotes the partial derivative of f with respect to its ith argument:

$$f_1(|\phi_i|, D_i) \equiv \frac{\partial f(|\phi_i|, D_i)}{\partial |\phi_i|}, \quad f_2(|\phi_i|, D_i) = \frac{\partial f(|\phi_i|, D_i)}{\partial D_i}. \tag{97}$$

We shall assume that

$$f_1(|\phi_i|, D_i) > 0, \quad f_2(|\phi_i|, D_i) < 0. \tag{98a,b}$$

The inequalities of (98) imply that diffusivities change faster at higher flux and that for a given flux the rate at which the diffusivities change is slower when the diffusivities themselves are larger.

Our mathematical model is completed by precise specification of the fact that $x_2 = \lambda$ is a source with a fixed flux ψ ($\psi > 0$), and $x_2 = 0$ is a sink at which the hormone concentration falls to zero:

$$\text{at } x_2 = \lambda, \ \phi_2 = \psi; \quad \text{at } x_2 = 0, \ s = 0. \tag{99a,b}$$

We now adopt our standard procedure of searching for steady-state solutions and then examining their stability. We expect a steady state in which there is a constant vertical flux of hormone, and no horizontal flux. Diffusivities in this simple case are expected to be constants. Thus, we search for a solution of the form

$$\phi_1(x_1, x_2, t) = 0 \quad \phi_2(x_1, x_2, t) = \psi,$$

$$D_1(x_1, x_2, t) = d_1, \quad D_2(x_1, x_2, t) = d_2, \tag{100}$$

where d_1 and d_2 are constants. Indeed, all conditions are satisfied if

$$s = \psi x_2/d_2, \quad f(0, d_1) = 0, \quad f(\psi, d_2) = 0. \tag{101a,b,c}$$

The latter two equations provide values for d_1 and d_2.

In the usual fashion, let us now introduce perturbations from the steady state, denoted by primes:

$$\phi_1 = 0 + \phi_1', \quad \phi_2 = \psi + \phi_2', \quad D_i = d_i + D_i', \quad s = (\psi x_2/d_2) + s'. \tag{102}$$

On linearization, the governing equations become [Exercise 12(a)]

$$\frac{\partial s'}{\partial t} = d_1 \frac{\partial^2 s'}{\partial x_1^2} + d_2 \frac{\partial^2 s'}{\partial x_2^2} + \frac{\psi}{d_2} \frac{\partial D_2'}{\partial x_2}, \tag{103a}$$

$$\frac{\partial D_1'}{\partial t} = u_1 |\phi_1'| + u_2 D_1', \tag{103b}$$

$$\frac{\partial D_2'}{\partial t} = v_1 \phi_2' + v_2 D_2', \tag{103c}$$

$$\phi_1' = d_1 \frac{\partial s'}{\partial x_1}, \quad \phi_2' = \frac{\psi}{d_2} D_2' + d_2 \frac{\partial s'}{\partial x_2}. \tag{103d,e}$$

Here, u_1, u_2, v_1, and v_2 are constants given by

$$u_1 = f_1(0, d_1), \quad u_2 = f_2(0, d_1), \quad v_1 = f_1(\psi, d_2), \quad v_2 = f_2(\psi, d_2). \tag{104}$$

The boundary conditions are

$$\text{at } x_2 = \lambda, \ \phi_2' = 0; \quad \text{at } x_2 = 0, \ s' = 0. \tag{105a,b}$$

In essence, our problem consists of the three linear equations (103a–c) for s', D_1', and D_2'. Once these functions are known, ϕ_1' and ϕ_2' can be determined from (103d,e).

As before, we attempt to solve these equations by suitable products of sines and cosines. Thus, we assume that the perturbation to the hormone concentration is given by

$$s'(x_1, x_2, t) = l \exp(pt) \cos(rx_1) \sin(qx_2), \tag{106}$$

where l, p, r, and q are constants. Note that the boundary condition (105b) is automatically satisfied. Substitution of (106) into (103a) yields

$$ps' = -d_1 r^2 s' - d_2 q^2 s' + \frac{\psi}{d_2} \frac{\partial D_2'}{\partial x_2}. \tag{107}$$

To satisfy this equation, we should choose D_2' so that its x_2 derivative has the same dependence on x_1, x_2, and t as does s'. Thus, we assume that

$$D_2'(x_1, x_2, t) = k \exp(pt) \cos(rx_1) \cos(qx_2), \tag{108}$$

where k is a constant. Substituting (108) into (107), we cancel the common factor of $\exp(pt) \cos(rx_1) \sin(qx_2)$ and obtain

$$pl = -d_1 r^2 l - d_2 q^2 l - (q\psi/d_2)k \tag{109}$$

from the requirement that (103a) be satisfied. Equation (103c) is now seen to require

$$pk = ak + v_1 d_2 ql, \quad \text{where } a \equiv v_2 + \psi v_1 / d_2. \tag{110a,b}$$

From (103b) we are led to

$$D_1'(x_1, x_2, t) = k' \exp(pt) |\sin(rx_1) \sin(qx_2)| \tag{111}$$

and [Exercise 12(b)]

$$pk' = u_1 d_1 rl + u_2 k'. \tag{112}$$

The remaining boundary condition (105a) requires that

$$\cos(q\lambda) = 0, \quad \text{i.e., } q = \frac{\pi}{2\lambda}, \, q = \frac{3\pi}{2\lambda}, \ldots. \tag{113}$$

Equations (109), (110), and (112) can be regarded as three linear equations for the three unknown constants l, k, and k'. The equations have the trivial solution $l = k = k' = 0$ corresponding to a zero perturbation. Equations (109) and (110) form a separate pair of equations for l and k. If the solution $l = k = 0$ is considered, then (112) requires

$$(p - u_2)k' = 0. \tag{114}$$

A nontrivial solution for k' is possible if and only if

$$p = u_2. \tag{115}$$

By eliminating one of the variables in the usual fashion, we easily find, in addition, that (109) and (110) can have a nontrivial solution only if [Exercise 12(c)]

$$p^2 + p(q^2 d_2 + r^2 d_1 - a) - (q^2 d_2 v_2 + a d_1 r^2) = 0. \tag{116}$$

To obtain information about stability, it remains to study the possible roots p of (115) and (116) – for then we know the behavior of the time factor $\exp(pt)$. One conclusion can be drawn at once: By (98b), the root of (115) always corresponds to a stable solution.

The simplest assumption we can make about the function f of (96) is that it is linear, so that

$$f(|\phi|, D) = \alpha|\phi| + \beta - \gamma D, \tag{117}$$

for certain constants α, β, and γ. By (98), α and γ are positive. Equation (101b) yields $d_1 = \beta/\gamma$. This shows that $\beta > 0$, for the steady-state diffusivity d_1 certainly must be positive. With this information, it is not difficult to show that the solution of (101) is stable for all permitted parameter values (Exercise 13).

In search of instability, we now consider a situation in which f has a quadratic dependence on the flux:

$$f(|\phi|, D) = \alpha|\phi|^2 + \beta - \gamma D. \tag{118}$$

By (101b,c),

$$\beta - \gamma d_1 = 0, \quad \alpha\psi^2 + \beta - \gamma d_2 = 0, \tag{119}$$

so that the steady-state diffusivities are given by

$$d_1 = \frac{\beta}{\gamma}, \quad d_2 = \frac{\alpha\psi^2 + \beta}{\gamma}. \tag{120}$$

Again $\alpha > 0$ and $\gamma > 0$, by (98), and $\beta > 0$ ensures the positivity of d_1.

We shall analyze equation (116) for the growth rate p under conditions (Exercise 15) such that

$$q^2 d_2 + r^2 d_1 - a > 0. \tag{121}$$

[If instability comes about by reversal of (121), then, by Exercise 15(a), growing oscillations begin. Monotonic instability somehow seems biologically more reasonable.]

Given (121), we know from our studies of the linear differential equation (A4.9) that instability will arise from (116) if and only if

$$ad_1 r^2 + d_2 v_2 q^2 > 0. \tag{122}$$

After a little manipulation [Exercise 14(a)], this condition can be written

$$q^2 < r^2(\beta/\alpha) \frac{\psi^2 - (\beta/\alpha)}{[\psi^2 + (\beta/\alpha)]^2}. \tag{123}$$

A necessary condition that makes it possible for the left side of (123) to be less than the right side is

$$\psi^2 > \beta/\alpha. \tag{124}$$

From (113), the parameter q is inversely proportional to the width λ of the tissue. We shall regard the instability as being caused by growth (i.e., by increase of λ). Equation (123) shows that the instability will first appear at the smallest permissible value of q, so that from the possibilities in (113) we select

$$q = \pi/2\lambda. \tag{125}$$

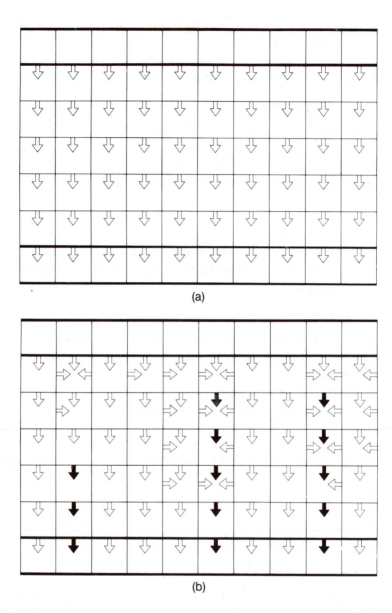

(a)

(b)

Figure 8.16. Simulation of a discrete version of (95), (96), and (118). (a) Original uniform flow between source (top) and sink (bottom). (b) Channeled pattern resulting from instability of (a). Shading of arrows indicates strength of flux, with black arrow denoting a flux 5–10 times as strong as that for white arrow. [From Mitchison (1977), *Science,* 196:270, Figure 4, which should be consulted for further details. Copyright 1977 by the AAAS.]

As λ increases and q decreases, condition (123) is satisfied more easily for large r. Note from (106) that $2\pi/r$ is the lateral periodicity of the perturbation. The smallest reasonable period is 2Δ, where Δ is the width of a cell. This would correspond to one file of cells bearing a diminished portion of flux, with a compensating surplus supplied by the file's immediate neighbors. We thus take

$$\frac{2\pi}{r} = 2\Delta, \quad r = \frac{\pi}{\Delta}. \tag{126}$$

Substituting (125) and (126) into (123), we obtain as the condition on the width that will yield instability

$$\lambda > \frac{\Delta}{2} \frac{\psi^2 + \beta/\alpha}{(\psi^2 - \beta/\alpha)^{1/2} (\beta/\alpha)^{1/2}}. \tag{127}$$

Having derived appropriate parameter ranges by linear stability theory, the next step is to perform computer simulations of the full set of equations. It is natural to employ a cellular version of the equations [compare (50)]. This was done by Mitchison (1980). Figure 8.16 shows how, under suitable conditions, an initially uniform flux becomes destabilized, resulting in a veinlike pattern. Note that although the order of magnitude is correct, simulation of the nonlinear equations gives a larger distance between "veins" than was predicted from linear theory. The considerations discussed in connection with (90) are, of course, relevant here too.

The papers of Mitchison (1980, 1981) should be consulted for further details, including discussion of experiments and examination of models that take into account known information concerning auxin transport.

Exercises

1. (a) Verify the following equations: (14), (16), (24), (33).
   (b) Show that as $U \to 0$ ($U > 0$), (42) is approximated by (43).

† 2. Consider the generalization of system (1) to the case in which y variation is also present. Let the definitions of (49) be generalized to

   $$a(i\Delta x, j\Delta y, t) \equiv a_{ij}(t),$$

   with a corresponding definition for $h_{ij}(t)$. What are the generalized versions of (50a) and (50b) in the case in which Δx and Δy are chosen to be equal?

† 3. (a) Given the parameter values listed in the legend for Figure 8.6, show that R is fairly large compared with unity and that $U \approx 0.027$, so that the approximate instability condition $U < U_c$ holds. Also verify the exact instability condition in the form $R > R_c$. Show that $L_c \approx 4.5\Delta x$.

†

(b) Using the data supplied in the legend for Figure 8.7, show that R is considerably larger than R_c [defined in (33)], so that the uniform state is quite strongly unstable. (Our analysis must be extensively revised if ρ_a and ρ_h vary. But to obtain approximate results, one can try replacing these quantities by their average values.)

(c) Calculate L_c for the cases of Figure 8.7 and Figure 8.13, and compare the predictions and the simulations.

4. (a) Verify (63) and (64a).

(b) Show directly that (20) and (54) have solutions of the form (65). Express the coefficients C_{01}, C_{02}, m_{01}, and m_{02} in terms of the quantities γ and β of (18), and thereby harmonize the present development with our earlier analysis of spatially uniform perturbations.

† 5. Derive equations of the form $A_{i1} + A_{i2} = p_i$, $C_{i1}A_{i1} + C_{i2}A_{i2} = q_i$, by multiplying both sides of (67a) and (67b) by $\cos(i\pi s/\lambda)$ and integrating from zero to λ. Although it is not always true, assume that the integral of the infinite sums is the sum of the individual integrals. Use a table of integrals, if necessary, to show that most terms in the sums are zero, and hence find explicit expressions for p_i and q_i in terms of integrals of $p(s)$ and $q(s)$.

6. Show that if $0 < U_c - U \ll 1$, then $m_{11} > 0$, but all the rest of the m_{ni} of (62) are negative.

7. (a) Verify equation (71).

(b) Discuss the remark in the text, below (71), on the relationship between the sign of A_{11} and the initial distributions of activator and inhibitor.

8. (a) Show that equation (19) is satisfied for the parameters of Figure 8.10.

(b) Verify equations (72) and (73).

9. (a) Verify equation (77).

(b) Show that the statement immediately following (77) is correct.

† (c) Show that maximizing m_{11} as a function of l^2 is equivalent to maximizing m_{11} as a function of l. Verify (81).

10. Verify equations (85), (86), and (89).

11. *Project:* Study the maximum of m_{11} in (78) as $U \to 0$ (a) for fixed P and (b) for fixed Q. Compare the predictions with (90) and with the simulation results. Keep in mind the possibilities that $l_M \to 0$ or $l_M \to \infty$ as $U \to 0$.

12. (a) Verify equation (103).

(b) Verify equations (107), (109), (110), (111), and (112).

(c) Verify equation (116).

13. Show that $a > 0$ is a necessary condition that (116) will yield an unstable root. Conclude that (117) cannot lead to instability.

14. (a) Derive equation (123).

(b) Show that equation (124) is equivalent to the condition $a > 0$ of Exercise 13.

15. (a) Show that if instability ensues by reversal of (121), then perturbation growth will be oscillatory.

(b) Equation (121) implies that $q^2 > d_2^{-1}(a - r^2 d_1)$. Thus, $d_2^{-1}(a - r^2 d_1)$ must be less than the right side of (123). What condition is thereby placed on the parameters? Is it consistent with (124)?

16. Show that if the concentration is fixed at $x_2 = \lambda$ [instead of (99a)], then the corresponding steady state is stable.

Supplement: Boundary conditions in the numerical analysis and model equations for nonlinear behavior

We have shown that the difference equations (8.50) can be taken as a discrete approximation to the differential equations (8.1) or, alternatively, as an independent model of reaction in well-mixed interiors of cells that are coupled by diffusion. We now discuss how the boundary conditions (8.53) are approximated in a computer analysis.

The simplest way to approximate the boundary condition $\partial a/\partial x = 0$ at $x = 0$ is to proceed as follows. For purposes of numerical analysis, the interval from 0 to L is divided up into N subintervals of width Δx by points at $x = 0, \Delta x, 2\Delta x, \ldots, (N-1)\Delta x, N\Delta x \equiv L$. An approximation to $\partial a/\partial x$ at $x = 0$ that uses this subdivision is

$$[a(\Delta x, t) - a(0, t)]/\Delta x. \tag{1}$$

Thus, to approximate $\partial a/\partial x = 0$, we simply set $a(\Delta x, t) = a(0, t)$, or, in the notation of (8.49),

$$a_1(t) = a_o(t). \tag{2a}$$

Similarly, we impose the conditions

$$h_1(t) = h_0(t), \quad a_{N-1}(t) = a_N(t), \quad h_{N-1}(t) = h_N(t). \tag{2b,c,d}$$

Let us carefully check to see if we have enough information to perform the desired calculations. Equation (8.50a) shows that calculation of $a_i(t + \Delta t) \equiv a(i\Delta x, t + \Delta t)$ requires information at time t, at the three adjacent points $(i-1)\Delta x$, $i\Delta x$, and $(i+1)\Delta x$. This is symbolized in Figure 8.17. Given the initial values of a and h, we can calculate all but two of the values of a at time Δt (Figure 8.18). The boundary conditions (2) supply the missing information. We can thus proceed step by step to calculate a and h at any time $p\Delta t$, $p = 1, 2, \ldots$.

If we take the point of view that the difference equations (8.50) are an independent model of the situation, we can argue as follows. We wish the cells at the end of our tissue to serve as a barrier to the flow of chemical. The flow of material between cells is assumed to be proportional to concentration difference. We thus place an imaginary cell next to each boundary cell and at each time step adjust the imaginary cell's concentration to match that of the adjacent boundary cell. This will ensure that indeed no chemical flows between the tissue and the boundary cell.

The "cellular" point of view can be expressed by equations (2a-d) if cells $i = 0$ and $i = N$ are regarded as "imaginary." If we were to take the

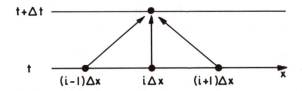

Figure 8.17. Diagrammatic representation of the fact that, according to (8.50), to calculate a and h at $(i\Delta x, t+\Delta t)$, we need information concerning a and h at $i\Delta x$ and the two adjacent points, at time t.

Figure 8.18. Representation of the fact that the scheme of Figure 8.17 gives a and h at the filled circles for $t=\Delta t$, given complete information at $t=0$. The boundary conditions (2) then allow us to fill in the missing information (open circles) that must be obtained before the next step forward in time can be made.

approach that we are approximating differential equations, we would include cells $i=0$ and $i=N$ in our results. We then would clearly see evidence of zero derivatives at the boundaries: The solution graphs would be horizontal at the two ends of the interval. Gierer and Meinhardt generally take the cellular point of view, and so omit the "imaginary" cells from their graphs. This explains, for example, why the graphs of the activator and inhibitor concentrations do not appear horizontal at the boundaries in Figure 8.10.

The basic ideas of numerical analysis are simple, but the subject contains pitfalls for the unwary and rewards for the subtle. Further study is advisable. N. Liron's contribution to the book edited by Segel (1980, Appendix 5) provides a brief discussion of numerical analysis aimed at a biological audience. The book by Richtmayer and Morton (1967) is the standard reference for numerical analysis of nonlinear diffusion equations.

As a final remark in connection with boundary conditions, we observe that our requirement that the boundaries be impermeable is just one possibility. Another important situation is that of a narrow ring of cells, in which case we would require

$$a(0)=a(L), \quad h(0)=h(L).$$

We can also require various influxes or leakages of chemicals through the boundaries.

Model equations for nonlinear behavior. We now present ordinary differential equations that illustrate the type of nonlinear behavior that was sketched in the section entitled "Further examination of pattern size."

Let us consider once again the standard case (8.56) in which U is just slightly below the critical value U_c, so that only the growth rate m_{11} of (8.66) is positive. For the moment, let us confine ourselves to solutions of spatial period L_c.

According to (8.71), under these circumstances linear theory shows that the activator perturbation, after a time, is proportional to $\exp(m_{11}\mu_a t) \times \cos(2\pi x/L_c)$. Let us denote by $A(t)$ the time-varying amplitude of the mode proportional to $\cos(2\pi x/L_c)$, and let us use a as an abbreviation for $m_{11}\mu_a$. Then the results of linear theory can be regarded as prescribing the equation

$$dA/dt = aA \tag{3}$$

for the amplitude function $A(t)$.

It turns out that (if a is sufficiently small) when nonlinear terms are taken into account, (3) becomes modified to

$$dA/dt = aA - a_1 A^3. \tag{4}$$

The somewhat lengthy calculations for the constant a_1 show that $a_1 > 0$ for the problem (8.1). The new term $-a_1 A^3$ represents the nonlinear *self-damping effect* to which we have referred. If we write (4) in the form

$$dA/dt = a(1 - a_1 a^{-1} A^2)A,$$

we can think of A as a population with birthrate $a(1 - a_1 a^{-1} A^2)$. For low population levels, the birthrate is approximately equal to a, but this rate decreases as the population increases. Equation (4) has a positive steady state $\sqrt{(a/a_1)}$ to which the population $A(t)$ tends as $t \to \infty$ [Exercises A1.10 and A5.8]. We thus see how a perturbation that grows exponentially when it is small eventually can approach a nonzero steady state (Figure 8.19).

It is a feature of linear theory that each of the various modes grows or decays independently of the other modes. Thus, in (8.61), every pair of equations for $\hat{A}_n(\tau)$ and $\hat{H}_n(\tau)$ is independent of all other pairs. When nonlinear effects are taken into account, however, all the modes are seen to interact. As a caricature of such an interaction, consider amplitudes $A(t)$ and $B(t)$ of two different modes. The spirit of what emerges from a proper nonlinear analysis is represented by the equations

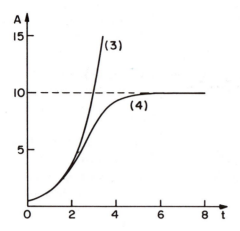

Figure 8.19. Graphs of typical solutions to (3) and (4) showing how nonlinear terms, although negligible for small time, eventually act to curb the initial exponential growth. Parameters: $a=1$, $a_1=0.01$, $A(0)=0.5$.

$$dA/dt = aA - a_1 A^3 - a_2 AB^2, \qquad (5a)$$

$$dB/dt = bB - b_1 BA^2 - b_2 B^3, \qquad (5b)$$

in which all coefficients are assumed positive (see Exercise A5.14). When $B=0$, equation (4) for A is recovered; when $A=0$, there is an entirely analogous equation for B. When both terms are present, however, the growth of each tends to diminish the "birthrate" of the other. For example, if (5a) is written

$$dA/dt = a(1 - a_1 a^{-1}A^2 - a_2 a^{-1}B^2)A,$$

we see that not only the growth of A but also the growth of B decreases the "birthrate" of mode A. This illustrates the concept that we have termed *intermode suppression*.

If

$$a_1/b_1 < a/b < a_2/b_2, \qquad (6)$$

then the phase portrait of the system (5) is given in Figure 8.20. We see that although both $A(t)$ and $B(t)$ grow exponentially when they are so small that nonlinear terms can be neglected, nonetheless either A or B ultimately "triumphs." Its competitor is driven to extinction. The decisive role of initial conditions is illustrated here, for the winner of the competition is determined by whether the initial point is located on one side or the other of the heavy trajectory in Figure 8.20. This situation is

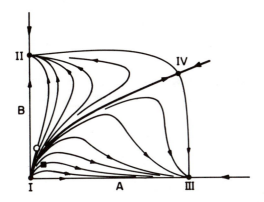

Figure 8.20. Schematic representation of the phase plane for system (5) subject to conditions (6), when $A \geqslant 0$, $B \geqslant 0$. Although both $A(t)$ and $B(t)$ grow exponentially when they are small, one ultimately becomes extinct. The filled square and the open circle typify initial conditions that result in the eventual dominance of A and B, respectively. Filled circles I–IV represent the steady-state solutions given in Exercise A5.14.

well exemplified in the competition between the two sinusoidal modes in Figure 8.11(b) and Figure 8.11(c).

Figure 8.20 is drawn for $a > b$. As we would expect when A and B are small, almost all initial points lead to a "victory" for A. Nonetheless, mode B can win out if its initial amplitude is sufficiently large compared with that of mode A.

Figure 8.21 depicts the phase plane when

$$a/b > a_2/b_2. \tag{7}$$

Here the intrinsic (no-competition) growth rate a of A is so much greater than that of B that A drives B to extinction whatever the initial value of B (except if A is initially zero). Thus, there is a limit to how much initial advantage can overcome intrinsic inferiority.

Note that the conditions (6), which lead to the behavior of Figure 8.20, necessarily imply

$$a_1 b_2 < b_1 a_2. \tag{8}$$

We see from the equations (5a,b) that the coefficients on the left of (8) represent self-damping, whereas those on the right represent intermode suppression. It is not surprising that intrinsic superiority can be entirely outweighed by initial advantage only when intermode suppression is sufficiently strong. "Coexistence" is possible if (8) is reversed, with both A and B tending to positive constants as $t \to \infty$.

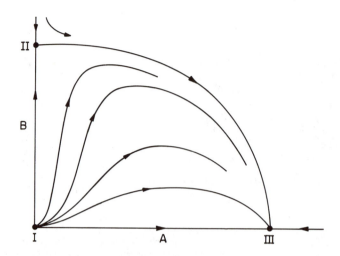

Figure 8.21. Phase plane for (5), as in Figure 8.20, except that the coefficients satisfy (7). Although both $A(t)$ and $B(t)$ grow initially, A always eventually drives B to extinction.

We have used a colorful vocabulary ("birthrate," "competition," "extinction," and "coexistence") in the first instance to render our exposition more attractive. But the fact of the matter is that the equations (5) can serve as a model for the interaction of two competitors. (In population dynamics, the nonlinear terms are more usually taken to be quadratic, but this does alter the conclusions to be drawn here.) Our discussion thus illustrates an important aspect of mathematical modeling: Sometimes, similar equations, and hence similar phenomena, occur in widely different fields. The modeling then plays an important unifying role, showing relationships where none were suspected, and allowing intuition from one field to be employed (cautiously) in another.

In the present case, the competition between the amplitudes of various sinusoidal modes is analogous to the competition between species of animals and plants. Matters such as nonlinear self-limitation and intermode (interspecies) suppression are common to these two entirely different contexts. Just as the struggle for existence results in the selection of a few species among many possibilities, so the nonlinear interaction between modes results in a characteristic distance between activator peaks, not a mere sum of all modes that can grow according to linear stability theory.

Not only species of plants and animals but also species of molecules have struggled for existence, with selection of a few from many possibilities. This is the line taken by Eigen and Shuster (1977, 1978) in a thoughtful attempt to explain the origin of life and the uniqueness of the genetic code. The mathematical setting is again that of systems of coupled ordinary differential equations, and again the nonlinear interactions serve to promote the existence of a few competitors at the price of the extinction of most.

9

A mechanical basis for morphogenesis

This chapter deals with the controlled generation of shape change (**morphogenesis**), a fundamental process in developmental biology. In its emphasis on the role of mechanical forces, the class of models presented here provides an instructive contrast to the completely chemical models discussed in Chapter 8. It is too early to say which of the two lines of attack is more fruitful, but is seems clear that both contend in an illuminating fashion with major phenomena in developmental biology.

Our presentation here is mainly based on papers by Odell, Oster, Alberch, and Burnside (1981) and Oster and Odell (1984). The necessity for relative brevity requires us to omit citation of the biological evidence for many of the assumptions. The original papers should be examined for this all-important matter, and for many other details that are omitted in the overall survey that will be given here.

We shall be concerned with phenomena such as gastrulation and neurulation, that is, with folding movements of cell sheets (epithelia). A schematic representation of such a sheet is given in Figure 9.1.

The key element in the theory is the finding that special contractile domains are located in the apical (top) region of the cells. When stretched a small amount and then released, these domains appear to behave like ordinary springs: They return to their undisturbed "rest length." But when the stretch is extended beyond a certain critical threshold, then on release the domains rapidly undergo a large contraction to a new and smaller rest length. On further contractions and releases, the cell always returns to its new rest length. Because the cell essentially retains its volume, the strong "purse-string" contraction causes the cell to become elongated (Figure 9.2).

Modeling active domains

Because the apical contractile domains are the key element of the phenomenon, we must examine their behavior in some detail. We note first

apical surface

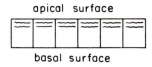

basal surface

Figure 9.1. Part of a cell sheet, with active apical domains denoted by wavy lines.

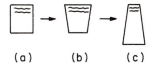

(a)   (b)   (c)

Figure 9.2. (a) Initial form of a cell. After a superthreshold extension (b), active fibers shrink to a new, shorter rest length (c). Because its volume is preserved, the cell becomes elongated.

that a sensible model for an isolated domain is the famous spring equation of elementary physics:

$$m\frac{d^2L}{dt^2} = -k(L-L_0) - \mu\frac{dL}{dt}. \tag{1}$$

Here, m represents the mass of the spring and $L(t)$ its length at time t, so that the left side of (1) constitutes the mass times the acceleration. By Newton's law, this must be equated to the sum of the forces that act on the spring. The first of these is an elastic force, proportional via the coefficient k to the extension from the rest length L_0. (We shall soon return to the question of representing the two possible rest lengths.) To this is added a damping force proportional to the velocity via the damping coefficient μ.

In the oozing, highly viscous motions of morphogenesis, the inertia term can be shown to be negligible (Odell et al., 1981). Thus, (1) can be approximated by

$$\frac{dL}{dt} = \frac{k}{\mu}(L_0 - L). \tag{2}$$

Think, for a moment, how we might build into a model the fact that the contractile filaments have two possible rest lengths. The idea of Odell and associates (1981) is to regard L_0 as a function of time and to postulate an equation for dL_0/dt in such a manner that the $L_0 - L$ phase plane will contain two stable steady states. A sketch of a phase plane with the desired properties is shown in Figure 9.3.

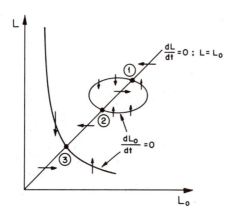

Figure 9.3. Horizontal and vertical nullclines that give three steady states (heavy dots) in the $L_0 - L$ phase plane. State 2 will turn out to be unstable.

For numerical computations, definite equations must be selected. The particular choice of Odell and associates (1981) is

$$\frac{dL_0}{dt} = \gamma(\epsilon_1^2 - LL_0)\left[\left(\frac{L-\eta}{\epsilon_3}\right)^2 + \left(\frac{L_0-\eta}{\epsilon_4}\right)^2 - 1\right] \tag{3}$$

where γ, ϵ_1, ϵ_3, ϵ_4, and η are constants, with

$$\eta = 1 - \epsilon_3\epsilon_4(\epsilon_3^2 + \epsilon_4^2)^{-1/2}. \tag{4}$$

It should be stressed that the qualitative features of the phase plane are important here, not their particular implementation. The cardinal feature [possessed by (2) and (3); see Exercise 1] is the existence of a **firing threshold,** the crossing of which engenders an essentially irreversible shift in the stable rest length from long to short (Figure 9.4).

One more detail is required to complete the model of the isolated active filament. To reproduce observations that the return to the contracted rest length is relatively rapid, the spring constant is taken to be larger when the rest length L_0 is smaller, by means of the assumption that

$$k = k_0/L_0, \quad k_0 \text{ a constant.} \tag{5}$$

Simulations of folding cell sheets

Having now a model for the key active element, it remains to represent the rest of the cell. Figure 9.5 diagrams the assumptions. In addition to

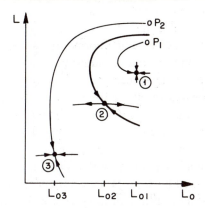

Figure 9.4. Further features of the phase plane of Figure 9.3. Steady states 1 and 3 are stable nodes; state 2 is a saddle point. The firing threshold is represented by the heavy trajectories entering state 2. As shown, after a subthreshold stretch from state 1 to P_1, the system will return to its initial state. After a superthreshold stretch to P_2, the system trajectory will approach state 3, with the rest length L_0 shifting from its previous long value L_{01} to a new reduced value L_{03}.

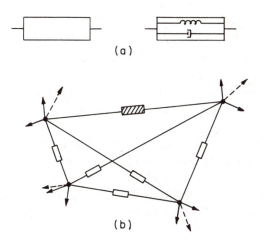

Figure 9.5. (a) The "white box" at left represents a spring-dashpot (viscoelastic) unit, as diagrammed at right. (b) Representation of the forces acting on a cell. Viscoelastic units act along the edges and diagonals, except that the apical surface contains a cross-hatched active element [which is governed by (2), supplemented by pressure and exterior forces, and (3)]. Neighboring cells exert forces at the vertices (dashed arrows). Pressure forces are represented by pairs of arrows at the vertices, normal to the edges.

the active element at the apex, the cell is endowed with ordinary damped springs along its edges and along its diagonals (to represent its viscoelastic interior). There is an internal pressure, acting normal to the boundaries, that very strongly resists compression. In addition, neighboring cells exert forces at the vertices of the given cell.

Equations involving the unknown intracellular forces are set up by requiring equilibrium conditions of vanishing force and moments at each cell vertex. (Remember that inertia is negligible, so that all forces must continually be in balance.) Most important are the equations for the active elements, which are differential equations of the form (2) and (3), except that (2) must be supplemented by appropriate components of the exterior forces arising from adjacent cells and of the pressure. The final mathematical problem turns out to be composed of $4N+4$ first-order ordinary differential equations, where N is the number of cells considered (typically about 30). These equations are, of course, solved numerically. On a fairly large computer (PDP-10), each numerical simulation required about half an hour of computer time.

Sample computer results are shown in Figure 9.6 for a simulation of amphibian neurulation. (The individual pictures are frames from a computer-generated movie.) The first frame depicts a cross section of the neural tube. A "prepattern" of lowered firing threshold is assumed to exist for the active elements in the cells that constitute the top half of the cell circle. If active elements in one or two cells at the top are initially made to cross their firing threshold, then, as shown, the rest of the sensitive cells rapidly contract to form the "neural plate." Later the plate rolls up into a tube.

Previously, authors have speculated on how some type of cellular "clock" could coordinate the cells to yield an orderly formation of the plate. The simulation shows how the coordination can be automatically accomplished by the appropriate mechanical laws. In addition, simulations can reproduce the experimental finding that an excised neural plate will fold into a tube.

Gastrulation in sea urchin is simulated in Figure 9.7. The chief additional feature here is an assumption that fluid in the central cavity (blastocoel) is extruded by elevated pressure.

Rather than present further illustrations of the model in action, we shall push on to explain mechanochemical ideas that might underlie the cytoplasmic contractility that has been postulated (Oster and Odell, 1984).

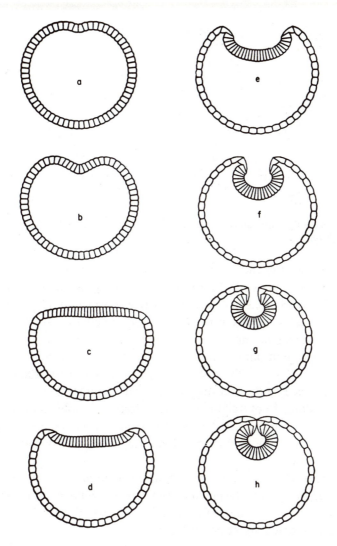

Figure 9.6. A simulation of amphibian neurulation; selected frames from a computer-generated movie. [From Odell et al. (1981), Figure 9.]

Calcium-actin-myosin interaction

Cell cytoplasm is now known to contain fibers of the protein actin. The fibers are cross-linked by further special proteins. If many cross-links are present, the cytoplasm is in a semisolid **gel** state; otherwise it is a liquid-like **sol**.

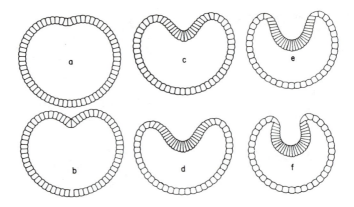

Figure 9.7. Simulation of sea urchin gastrulation. [From Odell et al. (1981), Figure 7.]

Calcium regulates the sol–gel balance by activating specific enzymes that cut the actin network. Moreover, Ca^{++} brings about activation of the protein myosin that in turn causes active contraction. (The contraction is presumably analogous to the well-studied actin-myosin "ratchet" cross-bridge action in smooth muscle, but the organized arrangements of actin and myosin in smooth muscle are absent in cytoplasm.)

The power of Ca^{++} to control cytoplasmic contractile behavior is greatly enhanced by the fact that calcium can stimulate its own release. How this is done is not well understood; a superthreshold concentration of Ca^{++} may trigger Ca^{++} release from internal stores, or a Ca^{++}-triggered contraction may bring about Ca^{++} release by mechanical means.

The properties of the calcium-modulated actin-myosin network will now be used to flesh out the previous highly phenomenological model for the active element. Our discussion will again be based on Newton's law (1), with the momentum neglected:

$$0 = -k(L, C)[L - L_0(C)] - \mu(C)\frac{dL}{dt}. \tag{6}$$

As before, the elastic force $-k(L - L_0)$ and the viscous force $-\mu(dL/dt)$ are postulated to be in equilibrium for an isolated element: Their sum is zero. In contrast to our earlier development, μ and L_0 will now be regarded as functions as C, and k will be taken to be a function of L and C.

The viscosity coefficient μ will be assumed to depend on the intracellular calcium concentration C in the manner depicted in Figure 9.8.

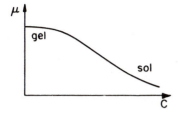

Figure 9.8. Dependence of viscosity on Ca concentration. The Jell-O-like highly cross linked material at low Ca is much more viscous than the liquidlike sol in which Ca-activated enzymes have made many cuts in actin fibers.

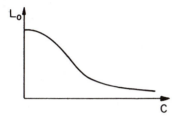

Figure 9.9. Postulated dependence of the rest length on the intracellular Ca concentration.

Resistance to shear (viscosity) is high at low values of C, for few actin-cutting enzymes are activated, and the cytoplasm will be in a highly cross-linked, resistive state. If more Ca^{++} is present (larger C), then the gel "melts," and the cytoplasm will be characterized by the smaller viscosity of the sol fluid.

The rest length L_0 at which the elastic force vanishes also depends on C, in the fashion shown in Figure 9.9. As the intracellular calcium concentration increases, contractile activity is enhanced, and the rest length decreases. The elasticity (or, more properly, the elastic modulus) k is expected to decrease as L increases, because there will be less overlap between fibers and hence fewer cross-links and active cross-bridges [Figure 9.10(a)]. The modulus k should at first increase when C increases, because of further Ca^{++}-activated contraction, but should eventually decrease when C is so high that the material becomes solated [Figure 9.10(b)].

In an isolated portion of cytoplasm, three factors are taken into account when considering the change in intracellular calcium concentration: (a) a release from internal stores at a rate proportional to the local

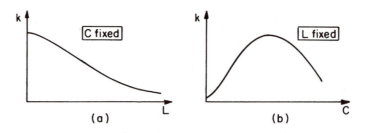

Figure 9.10. Expected dependence of the elasticity coefficient k on the amount of stretch (a) and on calcium (b).

"stretch" L; (b) a linear "disappearance" or resequestration; (c) a nonlinear, saturating autocatalytic internal release. These factors are represented in the specific equation

$$\frac{dC}{dt} = \gamma L - \nu C + \frac{\alpha C^2}{\beta + C^2}, \tag{7}$$

where γ, ν, α, and β are positive constants. To this equation we couple (6) in the form

$$\frac{dL}{dt} = \frac{k(L,C)}{\mu(C)}[L_0(C) - L]. \tag{8}$$

Effect of Ca-trigger maturation

Equations (7) and (8) describe the interaction between intracellular calcium C and the "stretch" L in an isolated active piece of cytoplasm. By phase-plane techniques, we now examine the qualitative behavior of the system (7) and (8) for various values of the parameter α. We assume that α slowly increases during the course of development. Because this parameter represents the maximum autocatalytic release, we can regard its increase as a maturing of the cell's Ca^{++}-trigger apparatus.

In the C–L phase plane, the two nullclines $dL/dt = 0$ and $dC/dt = 0$ are given by

$$L = L_0(C), \quad L = \frac{1}{\gamma}\left(\nu C - \frac{\alpha C^2}{\beta + C^2}\right). \tag{9a,b}$$

The intersections of these two curves give the steady-state points. The various possible cases are depicted in Figure 9.11. It can be shown that if the slope of the S-shaped nullcline of (9b) is positive at a steady-state

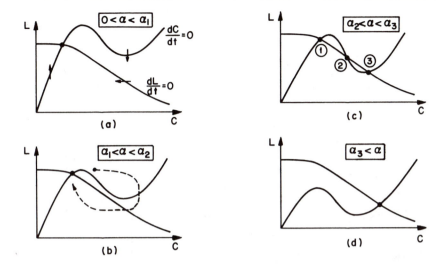

Figure 9.11. Qualitative plots of the nullclines (9) of the system (7)–(8) for increasing values of the trigger parameter α of (7). Heavy dots indicate steady states. The constants α_1, α_2, and α_3 denote transition values of α. (a) Single "relaxed" steady state (which turns out to be stable). (b) Excitable state. Although all trajectories eventually return to the single stable steady state, a trajectory that begins to the right of the hump in the vertical nullcline will make a large excursion in the phase plane before finally approaching the steady state. Compare Figure 6.13. (c) Three steady states: bistability. (d) Single contracted stable steady state.

point, then that point is stable. If the slope is sufficiently negative, the steady state is unstable (Exercise 2).

Contrast Figure 9.11(a) and Figure 9.11(d). In the first case, when α is small there is a stable equilibrium at a relatively large value of L, whereas in case (d), when α is large there is a stable equilibrium at which L is small. Thus, the passage of the system to a contracted steady state can be brought about automatically by a maturing of the Ca^{++}-trigger machinery.

In the intermediate situation of Figure 9.11(c), three steady states exist. Of these, two are stable, and the other is unstable. As in the situation of Figure 9.4, there is bistability, with a firing threshold permitting sufficient stretch to shift the rest length from long to short.[1]

1 In the particular model under consideration, the bistable state can be of two types. For $\alpha_2 < \alpha < \alpha_2'$, the situation is qualitatively the same as in Figure 9.4. For $\alpha_2' < \alpha < \alpha_3$, state 1 becomes an unstable spiral. Now the original system can exhibit oscillations, approaching a trajectory that circles steady-state point 1. Again, the system tends to one of two final states – a steady contracted state, as before, or an oscillatory relaxed state (Exercise 3).

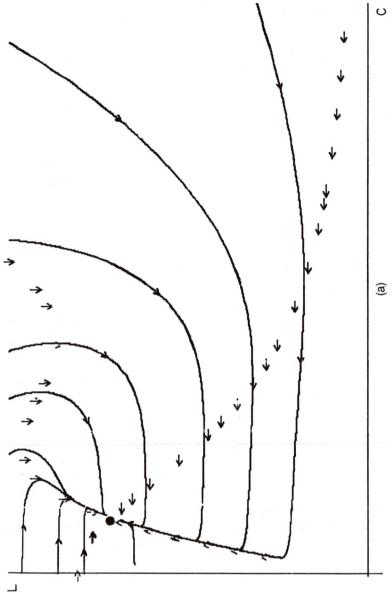

Figure 9.12. Computer-generated phase-plane trajectories (by G. Odell) corresponding to the various cases of Figure 9.11. (a) $\alpha = 0.33$. (b) $\alpha = 0.384$. (c) $\alpha = 0.4$. (d) $\alpha = 0.51$. Points on the nullclines are indicated by vertical and horizontal arrows. Note in (c) and (d) that the vertical nullcline drops below the C axis. This was not correctly depicted in Figure 9.11(c) and (d), but it is not germane to the qualitative features that are illustrated in Figure 9.11.

(a)

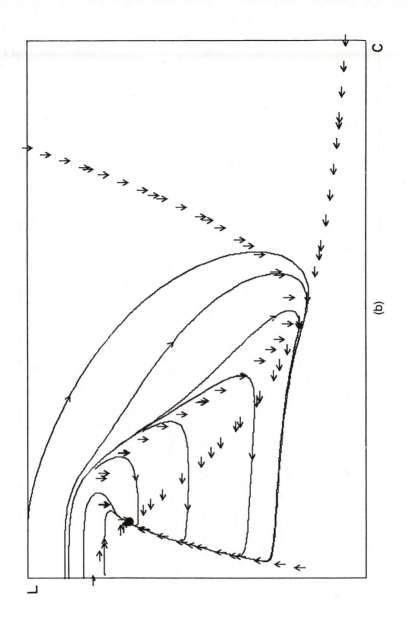

(b)

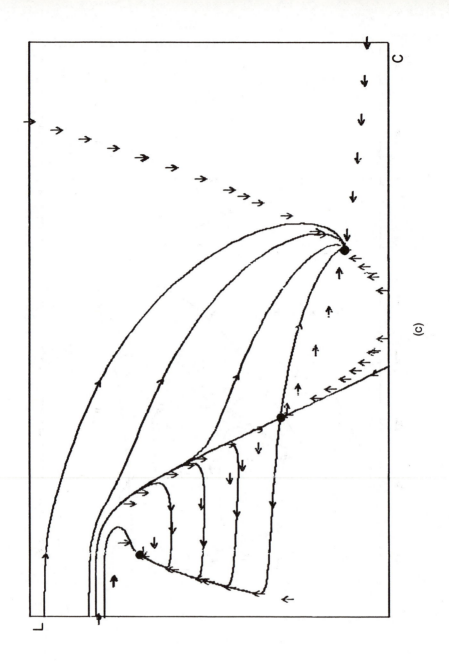

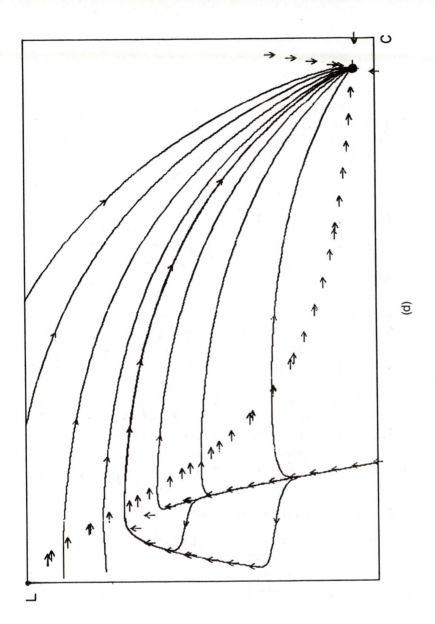

(d)

Before the bistable state is reached, there is **excitable** behavior, corresponding to Figure 9.11(b). Here there is a single stable steady state, but sufficient perturbation of the system can bring about a rather strong contraction before the system returns to its relaxed final state.

Figure 9.12 depicts computer-drawn collections of phase-plane trajectories corresponding to the four cases of Figure 9.11. The expected qualitative behavior is verified.

The excitability of Figure 9.11(b) and Figure 9.12(b) can explain the phenomenon of **focal contraction,** in which transient cell contractions are observed before neurulation commences. Figure 9.13(a) shows a sequence of tracings that depict a focal contraction in a *Taricha tarosa* embryo. Figure 9.13(b) quantitates the changes of one cell's circumference with time.

Equations (7) and (8) are employed to model a focal contraction. The right side of the force balance equation (8) is supplemented by the term $-\sigma/\mu(C)$, with σ a constant. This represents the effect of surrounding cells by a uniform tension σ. The computer simulation is started with the intracellular Ca concentration C set somewhat above its steady-state value, to simulate the effect of a spontaneous local influx of Ca. Parameters are chosen to give an excitable response. Figure 9.14(a) shows the excitable response in the phase plane for a particular parameter set. Figure 9.14(b) affirms that this model can mimic quite closely the principal features of the observations (assuming that the length of the region containing active fibers is proportional to the cell circumference).

The possible role of a developmental path

Our theoretical developments reveal a strong analogy between the behavior of the Ca^{++}-stimulated contractile apparatus on the one hand and the cAMP secretion machinery that was studied in Chapter 6. In both cases there is a **developmental path** wherein the slow change of one or more critical parameters switches behavior from one domain into another (Figure 9.15). In the present instance, we now show that because development along the path can proceed at different rates in different cells, we can provide a deeper understanding of how events such as neurulation (Figure 9.6) can occur.

Embryos are not uniform. (In general, there are already readily detectable gradients in the egg.) Suppose that there is a gradient in the rapidity of trigger machinery maturation, with more rapid maturation toward the top of Figure 9.6(a). One can easily imagine a situation in which at some

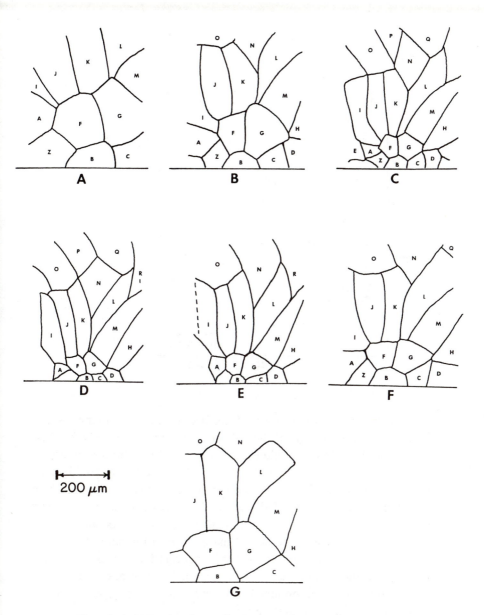

Figure 9.13. (a) Tracings from a film by A. Jacobson showing cells undergoing a focal contraction. Frames taken at $t = 0$, 6, 13, 19, 27, 40, and 68 min. (b) Circumference of cell G as a function of time. [From Oster and Odell (1984), Figures 8(a) and 8(b).]

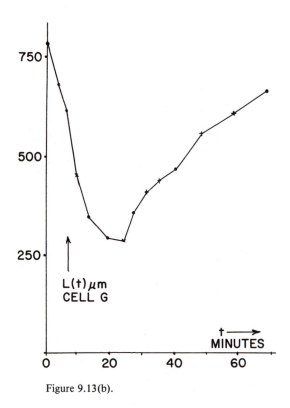

Figure 9.13(b).

time the parameter α has exceeded the critical value α_2, giving bistable behavior, for the top half of the cells in Figure 9.6(a) and only for these cells. Soon thereafter, α could exceed the second threshold α_3 in one or two of the topmost cells (compare Figure 9.15). These cells would immediately contract strongly, which in turn would induce a superthreshold stretch in the remaining top half of the cells. Further development would proceed as in Figure 9.6. Thus, mechanochemistry can operate to translate a simple gradient in the rate of maturation of the parameter α into the coordinated folding that characterizes neurulation. The mechanism is robust (immune to non-large disturbances), because almost the same final folding will be obtained for any reasonable type of polar-equatorial gradient in α maturation.

Chemical versus mechanical signaling

Until now we have confined our discussion of the Ca^{++}-controlled contractions to a model of the active element alone. Further progress

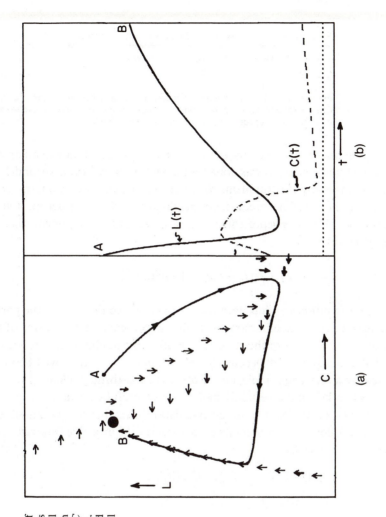

Figure 9.14. (a) Computer drawing of vertical and horizontal nullclines, plus excitable trajectory; for model of focal contraction, see text. (b) Plots of stretch L and intracellular Ca^{++} concentration C corresponding to the trajectory in (a). Note correspondence with data depicted in Figure 9.13(b). [From Oster and Odell (1984), Figure 8(c).]

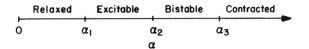

Figure 9.15. Different qualitative behaviors as a function of the maximum release parameter α. (The location of α_1 depends on how large an excursion in the phase plane is required to constitute an excitation.)

requires incorporating this element into a system of intracellular and intercellular forces, as was done for the earlier, more primitive model. As before, this requires supplementing the force equation with a term representing the exterior forces, from neighboring cells, that act on a given element. Thus, if σ represents the appropriate force component, then (6) is generalized to

$$\sigma_i = -k(L_i, C_i)[L_i - L_0(C_i)] - \mu(C_i)\frac{dL_i}{dt}. \tag{10}$$

Here, the subscript i has been used throughout to distinguish the particular cell under examination. In addition, we must take account of the flows J of Ca^{++} into the ith cell from its two neighbors. This is a somewhat delicate matter, because the elevation of Ca^{++} may well be mediated by a messenger molecule. (Pure diffusion through channels can be shown to be almost certainly ineffective.) Careful consideration (G. M. Odell and G. F. Oster, unpublished data) shows that it is permissible to use the following approximate expression for the effective flow rate from cell $i-1$ into cell i:

$$J_{i-1\to i} = \Gamma(C_{i-1} - C_i), \quad C_{i-1} > C_i,$$
$$= 0, \quad C_{i-1} \leqslant C_i. \tag{11}$$

(Here, Γ is a constant.) In any case, (7) must be generalized to

$$\frac{dC_i}{dt} = \gamma L_i + J_{i-1\to i} + J_{i+1\to i} - \nu C_i + \frac{\alpha_i C_i^2}{\beta + C_i^2}. \tag{12}$$

A major difference between chemical and mechanical coupling is seen by comparing Figure 9.6 with Figure 9.16. In the former, coupling was mechanical: γ was positive in (12), but the flow terms were negligible. In

Figure 9.16. A simulation of neurulation, as in Figure 9.6, except that intercellular communication is by the passage of a triggering chemical, not by stretch activation [i.e., $\gamma = 0$ in (12), but the flux terms J are large]. [Courtesy of G. M. Odell and G. F. Oster.]

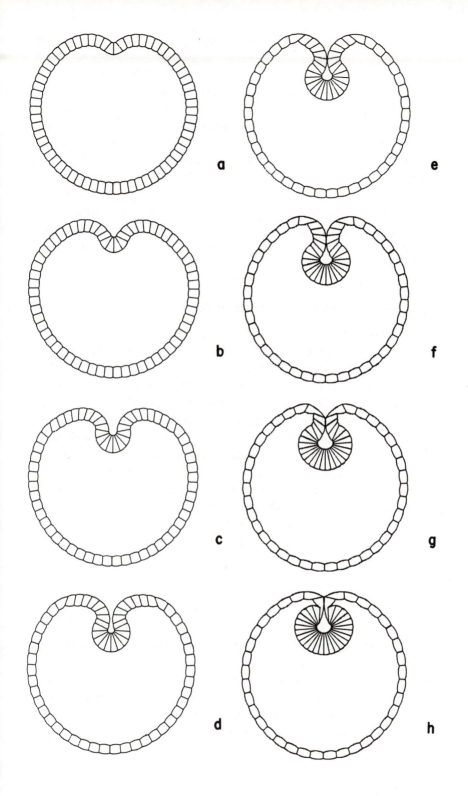

a

e

b

f

c

g

d

h

the latter, $\gamma = 0$ in (12), and intercellular Ca^{++} flow was strong. In contrast to the nearly simultaneous and parallel cellular action in the formation of the plate in Figure 9.6, Figure 9.16 shows a sequence of cellular contractions that are confined to the area of the invagination. Oster and Odell (unpublished data) remarked that "in computer animated movies of this simulation this localization is striking: cells seem to roll, one cell after another, around the fixed geometry of the lip of the invagination." It seems legitimate to speculate that chemical coupling is responsible for short-range coordination, whereas movements that are coordinated over many cells are likely to be associated with mechanical coupling. (Only electrical effects appear likely to be as capable of fast "long-range" information transfer.)

Skin organ primordia and cartilage condensations

We mention briefly here another major area of application of stability theory to morphogenesis via the interaction of mechanical forces. Murray, Oster, and Harris (1983) and Oster, Murray, and Harris (1984) formulated and analyzed a system of partial differential equations that describes cells migrating within an elastic matrix, producing traction on that matrix, and in turn being influenced by the resulting strain. Stability analysis and numerical solutions revealed various condensation phenomena that bear striking similarities, for example, to early events in feather formation and in cartilage condensation in developing limbs.

Harris, Stopak, and Warner (1983) experimentally demonstrated the existence of instabilities of the type considered. They showed that fibroblasts placed in homogeneous gels of collagen can form regular spatial arrays of cell and matrix condensations.

A final word

Several purposes have motivated the selection of material for this chapter. The dominant idea of qualitative behavior clued by stability calculations has been illustrated in another major context. Material on development in Chapter 8, where the emphasis was on the establishment of chemical pattern, has been contrasted with another approach that stresses the role of mechanical forces. In addition, the concept of a developmental path through zones of qualitative behavior, introduced in Chapter 6, has found further application.

We have repeatedly seen how systems of equations relevant to biology typically have several different types of qualitative behavior, with the type being fixed by the particular values taken by the various parameters that are important to the phenomenon under investigation. Zones of parameter values correspond to the same type of behavior. Transitions between zones can result from slow changes in parameters along some trajectories in parameter space.

There are signs of a revival of interest in the intimate connections between development and evolution. Transitions to a new spatial or temporal behavior via slow parameter changes have clear relevance to the genesis of new evolutionary types. A generalization of the developmental path called the "ontogenetic trajectory" has been proposed as an organizing principle for evolutionary transformations of form (Oster et al., 1980). It appears that awareness is growing of possible major roles that subtle dynamic phenomena may play in biology.

Exercises

1. Show that Figures 9.3 and 9.4 correctly represent the phase-plane behavior of equations (2) and (3).
2. Consider the system of equations

$$\frac{dC}{dt} = F(L, C), \quad \frac{dL}{dt} = G(L, C),$$

where [as in (7) and (8)]

$$F(L, C) = \gamma L - \nu C + \alpha S(C), \quad G(L, C) = k(L, C)[L_0(C) - L]/\mu(C).$$

Here, $S(C)$ is a saturating function satisfying $S(0) = 0$, $dS/dC \geq 0$, for $S \geq 0$, $S(C) \to 1$ as $C \to \infty$.

(a) Examine the conditions on $S(C)$ that are required to give the S-shaped nullcline $F = 0$ that is assumed in Figure 9.11. In particular, show that a Michaelean assumption for $S(C)$ [$S \sim C/(K+C)$] will not do the job.

(b) Show that perturbations C' and L' from steady states $\bar{C}$ and $\bar{L}$ satisfy

$$\frac{dC'}{dt} = F_C C' + \gamma L', \quad \frac{dL'}{dt} = QC' - PL',$$

where

$$F_C = -\nu + \alpha \frac{dS(C)}{dC}\Big|_{C=\bar{C}}, \quad Q = \frac{k(\bar{L}, \bar{C})}{\mu(\bar{C})}\left[\frac{dL_0(C)}{dC}\right]_{C=\bar{C}},$$

$$P = \frac{k(\bar{L}, \bar{C})}{\mu(\bar{C})}.$$

(c) Demonstrate that a steady state is stable if the corresponding value of F_C is negative. By implicitly differentiating $F(L, C) = 0$, show that in

this situation the slope of the humped nullcline is positive at the intersection that yields the steady state.

† 3. Sketch a possible phase portrait for the situation of Figure 9.11(c),
wherein steady state 1 is an unstable focus and is circled by a closed
curve representing a stable periodic solution (limit cycle).

The calculations that are relevant to cells migrating within an elastic
matrix require treating rather complicated equations, but something of
their spirit is shown in the next exercise (which assumes a knowledge of
the basic material in Chapter 8).

4. (a) The following equations have been proposed by Keller and Segel
(1970) as a very simple model for the response of slime mold cells
(density a) to the cAMP (density ρ) that they secrete:

$$\frac{\partial a}{\partial t} = \mu \frac{\partial^2 a}{\partial x^2} - \frac{\partial}{\partial x}\left(\chi a \frac{\partial \rho}{\partial x}\right), \quad \frac{\partial \rho}{\partial t} = fa - k\rho + D\frac{\partial^2 \rho}{\partial x^2}. \tag{13}$$

(Here we shall assume that μ, χ, f, k, and D are positive constants.)
Discuss in some detail what has been assumed, using the background
provided in Chapter 6.

(b) Show that there is a steady, spatially homogeneous solution $\bar{a}, \bar{\rho}$ if

$$f\bar{a} = k\bar{\rho}. \tag{14}$$

What is the intuitive reason for condition (14)?

(c) Examine the stability of the solutions of (b) to spatially periodic perturbations that are proportional to $\cos qx$. Show that instability will
occur if

$$(k + Dq^2) < \chi \bar{a} f/\mu.$$

(d) Analyze the consequences of (c). Then compare your conclusions
with those of Keller and Segel (1970), or of Segel (1980, Section 6.5).

Appendix 1

Mathematical prerequisites

We list here the mathematical ideas that are regarded as prerequisites for this course. Students can test their recall of the material by trying the exercises, for which answers are provided. Any standard calculus text will serve as a further reference. *Quick Calculus*, by Kleppner and Ramsay (1965), is a particularly brief text. The texts by Newby (1980) and DeSapio (1978) provide good introductions to calculus and are oriented to the biological sciences.

Function

The quantity y is a function of x for x in the interval (a, b); that is,

$$y = f(x), \quad a < x < b, \tag{1}$$

if for every x in the interval there is a rule allowing us to compute precisely one (finite) number y. Probably the single most important example is the linear function, whose graph rises at a relative rate given by its slope (Figure A1.1).

Limit

Suppose that $f(x)$ can be made as close as desired to L by choosing x sufficiently close to a. Then we write

$$\lim_{x \to a} f(x) = L. \tag{2}$$

Derivative

The derivative of $f(x)$ at $x = a$ is defined by

$$\left. \frac{df(x)}{dx} \right|_{x=a} = \lim_{h \to 0} \frac{f(a+h) - f(a)}{h}. \tag{3}$$

223

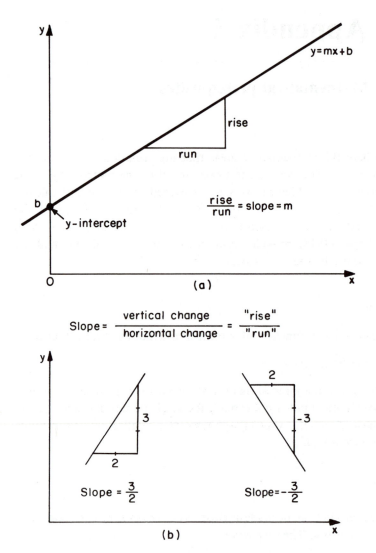

Figure A1.1. (a) The graph of a straight line. (b) Definition of slope, with examples.

The derivative at $x=a$ provides the slope of the tangent to $y=f(x)$ at $x=a$ (Figure A1.2).

$$\text{If } \frac{df}{dx} \text{ is } \begin{matrix} \text{positive} \\ \text{negative,} \end{matrix} \text{ then } f \text{ is } \begin{matrix} \text{increasing} \\ \text{decreasing.} \end{matrix} \qquad (4)$$

If $df/dx=0$ at $x=a$, then the function has a horizontal tangent at $x=a$,

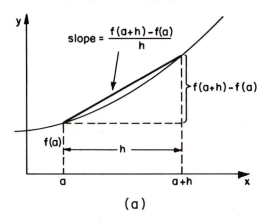

(a)

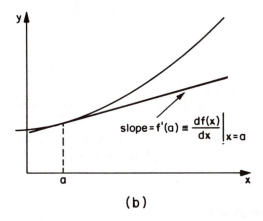

(b)

Figure A1.2. (a) The derivative of $f(x)$ at $x=a$ is the limit of the slope of the secant (heavy line) as h approaches zero. (b) The derivative at a point gives the slope of the tangent line at that point.

and the function at that point has a maximum or a minimum or a point of inflection (Figure A1.3).

Higher derivatives

The second derivative is the derivative of the first derivative, and so forth:

$$\frac{d^2 f}{dx^2} = \frac{d}{dx}\left(\frac{df}{dx}\right), \quad \frac{d^3 f}{dx^3} = \frac{d}{dx}\left(\frac{d^2 f}{dx^2}\right). \tag{5}$$

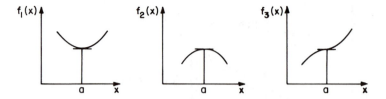

Figure A1.3. If the derivative is zero at $x=a$, at that point the function may have (a) a minimum, (b) a maximum, or (c) a point of inflection.

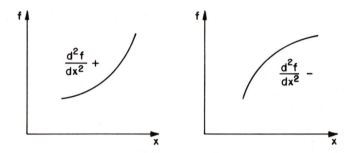

Figure A1.4. If the second derivative is positive (negative), the curve is concave upward (downward). A mnemonic is "plus, holds water – minus, spills water."

If the second derivative is positive (negative), then the first derivative is increasing (decreasing), so that the graph is concave upward (concave downward); see Figure A1.4.

Basic rules for manipulating derivatives

The derivative of the sum (difference) is the sum (difference) of the derivatives:

$$\frac{d}{dx}(f+g)=\frac{df}{dx}+\frac{dg}{dx}, \quad \frac{d}{dx}(f-g)=\frac{df}{dx}-\frac{dg}{dx}. \tag{6}$$

The derivative of a constant times a function equals the constant times the derivative:

$$\frac{d}{dx}(Cf)=C\frac{df}{dx}. \tag{7}$$

The derivative of a product of two functions is the first times the derivative of the second plus the second times the derivative of the first:

$$\frac{d}{dx}(fg)=f\frac{dg}{dx}+g\frac{df}{dx}. \tag{8}$$

The derivative of the quotient of two functions is the denominator times the derivative of the numerator minus the numerator times the derivative of the denominator, all divided by the denominator squared:

◆◆
$$\frac{d}{dx}\left(\frac{f}{g}\right) = \frac{g(df/dx) - f(dg/dx)}{g^2}.$$
(9)

The chain rule

If f is a function of x and x is a function of t, then the derivative of f with respect to t is the derivative of f with respect to x times the derivative of x with respect to t:

◆◆
$$\frac{df}{dt} = \frac{df}{dx}\frac{dx}{dt}, \quad \text{or, more precisely,} \quad \frac{d}{dt}f[x(t)] = \frac{df(x)}{dx}\bigg|_{x=x(t)} \frac{dx(t)}{dt}.$$
(10)

Implicit differentiation

If a function is given implicitly by an equation, then an expression for the derivative can be found by differentiating each term in the equation. For example,

$$\text{if } x^2 + [f(x)]^3 = x+1, \text{ then } 2x + 3f^2\frac{df}{dx} = 1,$$
(11)

yielding

$$\frac{df}{dx} = \frac{1-2x}{3f^2}.$$

Important functions and their derivatives

The derivative of a quantity raised to some power is given by the rule "power down, power diminished by one":

◆◆
$$\frac{dx^n}{dx} = nx^{n-1}.$$
(12)

Trigonometric functions (Figure A1.5) have simple derivative formulas if the angles are measured in radians. (A right angle equals $\pi/2$ radians; other angles are in proportion, e.g., $45° = \pi/4$ radians.)

◆◆
$$\frac{d}{dx}\sin x = \cos x, \quad \frac{d}{dx}\cos x = -\sin x.$$
(13)

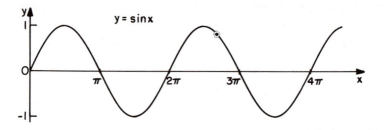

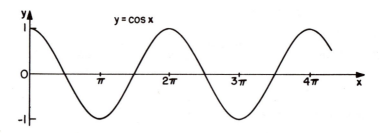

Figure A1.5. Graphs of sin x and cos x.

The remaining trigonometric functions are related to sin and cos by the formulas

$$\tan x = \sin x/\cos x, \quad \cot x = 1/\tan x, \quad \csc x = 1/\sin x, \quad \sec x = 1/\cos x. \tag{14}$$

There is no need to remember the derivatives of the functions defined in (14); we need only apply (13) and the differentiation rules. For example:

$$\frac{d}{dx}(\tan x) = \frac{d}{dx}\left(\frac{\sin x}{\cos x}\right) = \frac{(\cos x)(\cos x) - (\sin x)(-\sin x)}{\cos^2 x}$$

$$= \frac{1}{\cos^2 x} = \sec^2 x.$$

The natural logarithm and the exponential function are inverses:

$$\ln(e^x) = x, \quad e^{\ln x} = x. \tag{15}$$

Their derivatives satisfy

$$\frac{d}{dx}\ln x = \frac{1}{x}, \quad \frac{d}{dx}e^x = e^x. \tag{16}$$

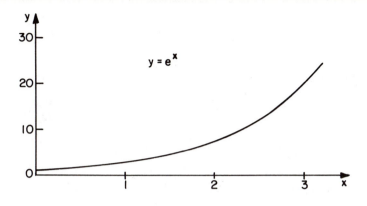

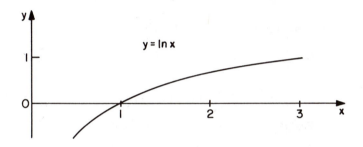

Figure A1.6. Graphs of e^x and $\ln x$ (note different vertical and horizontal scales).

Recall that

♦♦ $$e^{x+y} = e^x e^y, \quad \ln(xy) = \ln x + \ln y, \tag{17}$$

♦♦ $$e^0 = 1, \quad \ln 1 = 0,$$

and that $\exp(x)$ is synonymous with e^x. See Figure A1.6.

Partial derivative

If a partial derivative is taken with respect to a certain variable, all other variables are regarded as constants. Examples:

$$\frac{\partial}{\partial x}(x^2 y) = 2xy, \quad \frac{\partial}{\partial t} \sin(zt^2 x) = 2tzx \cos(zt^2 x).$$

For most functions, the order of partial differentiation is immaterial. For example:

$$\frac{\partial}{\partial y}\frac{\partial}{\partial x}(x^2y) \equiv \frac{\partial^2}{\partial y \partial x}(x^2y) = 2x, \quad \frac{\partial}{\partial y}(x^2y) = x^2, \quad \frac{\partial^2}{\partial x \partial y}(x^2y) = 2x.$$

Exercises

† 1. (a) $\dfrac{d}{dx}(x^3)$. (b) $\dfrac{d}{dx}(x^{-2})$. (c) $\dfrac{d}{dx}\left(\dfrac{1}{x^5}\right)$. (d) $\dfrac{d}{dx}(x^2 + x^{3.5})$.

(e) $\dfrac{d}{dx}(7x^{-3})$. (f) $\dfrac{d^2}{dx^2}(x^4)$. (g) $\dfrac{d^3}{dx^3}(x^{-1})$. (h) $\dfrac{d}{dx}(x^{-2} - x^3)$.

† 2. (a) $\dfrac{d}{dx}(x \sin x)$. (b) $\dfrac{d}{dx}\left(\dfrac{x}{\sin x}\right)$.

(c) $\dfrac{d}{dx}\cos(2x)$. (d) $\dfrac{d}{dx}\cos(x^2)$.

† 3. Using equation (14), find the following:

(a) $\dfrac{d}{dx}\sec x$. (b) $\dfrac{d}{dx}\csc x$. (c) $\dfrac{d}{dx}\tan x$. (d) $\dfrac{d}{dx}\cot x$.

† 4. (a) $\dfrac{d}{dx}e^{x^2}$. (b) $\dfrac{d}{dx}\ln(\sin x)$. (c) $\dfrac{d}{dx}\dfrac{\ln x}{e^x}$. (d) $\dfrac{d}{dx}\exp(5x^3)$.

† 5. Find dy/dx if x and y are related as follows:

(a) $x^2 + y^2 = 25$. (b) $xe^y = 3$. (c) $\ln(x+y) = x^2y$. (d) $\sin 3x = \cos 2y$.

6. †(a) $\dfrac{\partial}{\partial x}(10x^3y^2)$. †(b) $\dfrac{\partial}{\partial z}(e^{xyz^2})$. †(c) $\dfrac{\partial}{\partial t}\sin(e^{zt})$.

(d) If $f(p,K) = (1 - Kp)/[2 - (1+K)p]$,

show that $\partial f/\partial p = (1 - K)/[2 - (1+K)p]^2$.

7. Verify that $P(t) \equiv P_0 \exp(-Rt)$ satisfies $dP/dt = -RP$, $P(0) = P_0$.

8. In density-gradient centrifugation, after a long time the concentration C of a macromolecular solute will reach a steady state determined by an equation that gives a balance between diffusion and centrifugal force. The equation for C as a function of position r is

$$\frac{dC}{dr} = -\omega^2 MGQC(r - r_0), \tag{18}$$

where ω is the angular velocity of the centrifuge, M is the molecular weight of the molecule, G is the given density gradient, and Q and r_0 are constants. Rubinow (1975, p. 255) gave as the solution of (18):

$$C = C(r_0) \exp[-\tfrac{1}{2}\omega^2 MGQ(r - r_0)^2]. \tag{19}$$

[Recall that $e^x \equiv \exp(x)$.] A student tries to check this result by seeing if the derivative of (19) is given by (18). After doing so, the student asserts that the factor $\tfrac{1}{2}$ in (19) should not appear. Who is right, the student or Rubinow?

9. According to Parnas and Segel (1980), under certain circumstances the facilitation (F) of neurotransmitter release depends on the amount E of calcium that enters the nerve terminal according to the formula

$$F = \frac{\theta(K+E)}{K+\theta E},$$

where θ and K are constants; $1 < \theta < 2$, $K > 0$.
(a) Show that dF/dE is negative.
(b) E depends on the concentration C of calcium outside the terminal. These authors state that ''as long as we make the assumption that E is an increasing function of C, then F is a decreasing function of C.'' Show that this is a correct statement by means of an equation involving dF/dC.
(c) Find dF/dE if ''cooperativity'' is present, so that

$$F = \left(\frac{\theta(K+E)}{K+\theta E}\right)^n.$$

10. (a) Show that

$$A(t) = e^{at}[c + (a_1/a)e^{2at}]^{-1/2} \tag{20}$$

satisfies the equation

$$dA/dt = aA - a_1 A^3, \tag{21}$$

where a, a_1, and c are positive constants. [Differential equation (21) is discussed in the supplement to Chapter 8.] The reader probably will find it helpful to introduce $x(t) \equiv e^{at}$, so that $x^2 = e^{2at}$.
(b) Give arguments in favor of the assertion that (20) implies

$$\lim_{t \to \infty} A(t) = (a/a_1)^{1/2}.$$

Appendix 2

Infinite series and Taylor approximations

An infinite series of constants

$$\sum_{n=0}^{\infty} c_n = c_0 + c_1 + c_2 + \dots$$

is said to **converge** to a sum S if the sum of the first $N+1$ terms comes arbitrarily close to S as N becomes larger and larger, that is, if

$$\lim_{N \to \infty} (c_0 + c_1 + \dots + c_N) = S. \tag{1}$$

An infinite series of functions $\sum_{n=0}^{\infty} C_n(x)$ converges to $S(x)$ for $a < x < b$ if for every constant x_0, $a < x_0 < b$, the infinite series of constants $\sum_{n=0}^{\infty} C_n(x_0)$ converges to the sum $S(x_0)$.

The most important series of functions are **power series**, also called **Taylor series**, where each term is a constant multiple of $(x-a)^n$ for a constant a and a nonnegative integer n. Suppose that

$$f(x) = \sum_{n=0}^{\infty} \alpha_n (x-a)^n; \quad \alpha_0, \alpha_1, \dots \text{ constants}; \quad |x-a| < R; \tag{2}$$

that is, suppose that the power series converges to $f(x)$ for $|x-a| < R$. Then, substitution of $x = a$ into

$$f(x) = \alpha_0 + \alpha_1(x-a) + \alpha_2(x-a)^2 + \dots$$

yields

$$f(x) = \alpha_0 + 0 + 0 + \dots.$$

Differentiating (2), we obtain

$$f'(x) = \sum_{n=1}^{\infty} n\alpha_n (x-a)^{n-1}, \quad ' \equiv d/dx,$$

or

$$f'(x) = \alpha_1 + 2\alpha_2(x-a) + 3\alpha_3(x-a)^2 + \dots.$$

Another substitution gives

232

$$f'(a) = \alpha_1 + 0 + 0 + \dots .$$

Similarly,

$$f''(x) = \sum_{n=2}^{\infty} n(n-1)\alpha_n (x-a)^{n-2}, \quad f''(a) = 2\alpha_2,$$

$$f'''(a) = 3 \cdot 2 \cdot \alpha_3, \quad f^{(4)}(a) = 4 \cdot 3 \cdot 2 \cdot \alpha_4.$$

In general,

$$\alpha_n = \frac{1}{n!} \frac{d^n f(x)}{dx^n}\bigg|_{x=a} \equiv \frac{f^{(n)}(a)}{n!}, \tag{3}$$

where we have used $f^{(n)}(a)$ as an abbreviation for the nth derivative of f, evaluated at $x=a$. The combination of (2) and (3) gives the Taylor-series formula

◆◆
$$f(x) = \sum_{n=0}^{\infty} \frac{f^{(n)}(a)(x-a)^n}{n!}. \tag{4}$$

Most often used is the special case wherein $a = 0$:

◆◆
$$f(x) = \sum_{n=0}^{\infty} \frac{f^{(n)}(0)}{n!} x^n. \tag{5}$$

Remarks. (a) It is shown in calculus texts that the power series (2) does indeed always converge, for some nonnegative R, in the interval

$$|x-a| < R, \quad \text{i.e.,} \quad -R < x-a < R.$$

It turns out that the **radius of convergence** R is given by

$$R = \lim_{n \to \infty} \left| \frac{\alpha_n}{\alpha_{n+1}} \right|, \tag{6}$$

if this limit exists. If $R = \infty$, the series converges for all finite x.

   (b) Differentiation of the infinite power series as if it were a polynomial, by summing the derivatives of each term, can be justified.

   (c) By definition, the more terms we take in a convergent power series, the better is the approximation to the function f given by the power series. The two-term approximation

$$f(x) = f(a) + (x-a)f'(a) \tag{7a}$$

is equivalent to approximating $f(x)$ near a by the tanget line at $x=a$ (Figure A2.1).

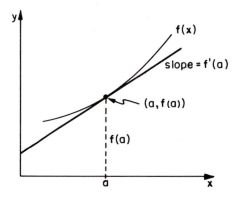

Figure A2.1. The heavy straight line is an approximation to $y=f(x)$ when $x-a$ is small. The slope of this line is $f'(a)$. Its equation is $y=f'(a)x+b$, where b must be chosen so that the point $[a,f(a)]$ is on the line. This requires $f(a)=f'(a)\cdot a+b$ or $b=f(a)-af'(a)$. Thus, the equation of the heavy straight line is $y=f'(a)x+[f(a)-af'(a)]=f(a)+(x-a)f'(a)$.

The **Taylor approximation** (7a) is often used in a slightly different form, obtained by introducing the change of variable

$$x-a=x', \quad \text{so that } x=a+x'.$$

If x is close to a, so that the **perturbation** x' is small, we can use (7a), which now takes the form

◆◆
$$f(a+x') \approx f(a)+f'(a)x', \quad f'(a) \equiv \frac{df(x)}{dx}\bigg|_{x=a}. \qquad (7b)$$

For functions of two variables, the Taylor approximation formula (7b) generalizes to

◆◆
$$f(a+x',b+y') \approx f(a,b)+f_x(a,b)x'+f_y(a,b)y', \qquad (8a)$$

where

$$f_x(a,\smile, \equiv \frac{\partial f(x,y)}{\partial x}\bigg|_{\substack{x=a\\y=b}}, \quad f_y(a,b) \equiv \frac{\partial f(x,y)}{\partial y}\bigg|_{\substack{x=a\\y=b}}. \qquad (8b)$$

The proof is quite simple – see Exercise 3.

Example: If

$$\frac{1}{1-x} = \alpha_0+\alpha_1 x+\alpha_2 x^2+\ldots = \sum_{n=0}^{\infty} \alpha_n x^n,$$

find the constants α_i.

Solution:

$$f(x)=(1-x)^{-1}, \quad f'(x)=(1-x)^{-2}, \quad f''(x)=2(1-x)^{-3}, \quad f'''(x)=6(1-x)^{-4}.$$

$$f(0)=1, \quad f'(0)=1, \quad f''(0)=2, \quad f'''(0)=3!, \quad f^{(n)}(0)=n!.$$

$$\alpha_n \equiv \frac{f^{(n)}(0)}{n!}=1.$$

Thus, we have derived the **geometric series**

$$(1-x)^{-1}=1+x+x^2+x^3+\ldots. \tag{9}$$

Because

$$\lim_{n \to \infty} \left| \frac{\alpha_n}{\alpha_{n+1}} \right| = \lim_{n \to \infty} 1 = 1.$$

we have $R=1$, and (9) converges for $|x|<1$.

Example: If

$$e^x = \sum_{n=0}^{\infty} \alpha_n x^n,$$

find the constants α_n and the radius of convergence.
Solution:

$$f(x)=e^x, \quad f'(x)=e^x, \quad f^{(n)}(x)=e^x, \quad f^{(n)}(0)=1.$$

$$\alpha_n = \frac{1}{n!}, \quad \lim_{n \to \infty} \left| \frac{\alpha_n}{\alpha_{n+1}} \right| = \lim_{n \to \infty} (n+1) = \infty.$$

Thus,

$$e^x = \sum_{n=0}^{\infty} \frac{x^n}{n!} \quad \text{for } |x|<\infty, \text{ i.e., for all } x. \tag{10}$$

Exercises

1. Derive the following Taylor-series formulas:

(a) $\sin x = x - \dfrac{x^3}{3!} + \dfrac{x^5}{5!} - \ldots;$ \hfill (11)

(b) $\cos x = 1 - \dfrac{x^2}{2!} + \dfrac{x^4}{4!} - \ldots;$ \hfill (12)

(c) $\ln(1+x) = x - \dfrac{x^2}{2} + \dfrac{x^3}{3} - \ldots.$ \hfill (13)

(d) $f(a+x) = \displaystyle\sum_{n=0}^{\infty} \dfrac{f^{(n)}(a)x^n}{n!}.$ \hfill (14)

†   2. (a) If

$$\sin x = a_0 + a_1\left(x - \frac{\pi}{2}\right) + a_2\left(x - \frac{\pi}{2}\right)^2 + \ldots,$$

find a_0, a_1, and a_2.

(b) If $f(x) = \ln(3 + x^2)$ and $0 < \epsilon \ll 1$, show that $f(1 + \epsilon) \approx \ln 4 + \frac{1}{2}\epsilon$, $f(2 + \epsilon) \approx \ln 7 + \frac{4}{7}\epsilon$.

3. Use appropriate versions of (7b) – and another approximation – to prove (8), starting with the identity

$$f(a + x', b + y') - f(a, b) = f(a + x', b + y') - f(a + x', b)$$
$$+ f(a + x', b) - f(a, b).$$

4. If $f(x, y) = x^2 y^3$, show that $f(1 + \epsilon, 2 + \delta) \approx 8 + 16\epsilon + 12\delta$ if $0 < \epsilon \ll 1$, $0 < \delta \ll 1$.

Appendix 3

Difference equations[1]

For definiteness, consider a situation (as in Chapter 2) in which N_{i+1}, the number of organisms per unit area during the year $i+1$, depends only on the number of organisms that existed the previous year:

$$N_{i+1} = f(N_i). \tag{1}$$

This equation is called a **difference equation**. It is **first-order**, because determination of N_{i+1} requires knowledge of only a single past generation. A specific example of a **second-order difference equation** is

$$N_{i+1} = N_i + N_{i-1}^2. \tag{2}$$

In general, we speak of **nth-order difference equations**.

For (2), we can compute all future values of N_{i+1}, N_{i+2} if, as **initial conditions**, the populations for two consecutive years are known. Thus, if

$$N_0 = 1, \quad N_1 = 2,$$

then, from (2),

$$N_2 = 2 + 1^2 = 3, \quad N_3 = 3 + 2^2 = 7, \quad \text{etc.}$$

For (1), a single initial condition

$$N_0 = C, \quad C \text{ given}, \tag{3}$$

is sufficient. Hand calculators, particularly programmable ones, are excellent aids for quickly finding a number of successive values N_i. Analytic methods are used to find general formulas, and hence to determine the influences of various parameters.

Linear difference equations

Equations of the form

$$a_i N_i + a_{i-1} N_{i-1} + \ldots + a_1 N_1 + a_0 N_0; \quad a_i, a_{i-1}, \ldots, a_0 \text{ known}, \tag{4}$$

1 The reader unfamiliar with complex numbers (Appendix 6) is advised on first reading to skip Example 2 and related material on certain second-order difference equations. Such equations appear only in Exercise 3.5.

are called **linear**. If the coefficients a_i do not depend on the index i, then the equation is said to have **constant coefficients**. Thus, for example, the following equations are, respectively, linear, linear with constant coefficients, and nonlinear:

$$7iN_i + 8N_{i-1} = (\sin i)N_{i-2}, \tag{5a}$$

$$16N_{i+3} - 2N_{i+2} + 3N_{i+1} - N_i = 0, \tag{5b}$$

$$N_{i+1}^2 = 3N_i. \tag{5c}$$

All equations called linear have the two fundamental properties (a) that a *constant multiple of a solution is a solution* and (b) that the *sum of two solutions is a solution*. To prove (a), assume that $N_i^{(1)}$ is a solution of (4), so that

$$a_i N_i^{(1)} + a_{i-1}N_{i-1}^{(1)} + \ldots + a_0 N_0^{(1)} = 0. \tag{6}$$

The constant multiple $CN_i^{(1)}$ is a solution if

$$a_i(CN_i^{(1)}) + a_{i-1}(CN_{i-1}^{(1)}) + \ldots + a_0(CN_0^{(1)}) = 0, \tag{7}$$

but the left side of (7) is equal to

$$C(a_i N_i^{(1)} + a_{i-1}N_{i-1}^{(1)} + \ldots + a_0 N_0^{(1)}) = 0.$$

To prove (b), suppose that $N_i^{(2)}$ is also a solution of (4), so that

$$a_i N_i^{(2)} + a_{i-1}N_{i-1}^{(2)} + \ldots + a_0 N_0^{(2)} = 0. \tag{8}$$

The quantity $N_i^{(1)} + N_i^{(2)}$ is also a solution if

$$a_i(N_i^{(1)} + N_i^{(2)}) + a_{i-1}(N_{i-1}^{(1)} + N_{i-1}^{(2)}) + \ldots + a_0(N_0^{(1)} + N_0^{(2)}) = 0. \tag{9}$$

But by rearrangement, the left side of (9) is equivalent to

$$(a_i N_i^{(1)} + a_{i-1}N_{i-1}^{(1)} + \ldots + a_0 N_0^{(1)}) + (a_i N_i^{(2)} + a_{i-1}N_{i-1}^{(2)} + \ldots + a_0 N_0^{(2)})$$

$$= 0 + 0 = 0.$$

Combining (a) and (b), we have the **superposition principle**, which states that if $N_i^{(1)}$ and $N_i^{(2)}$ are solutions of (4), then so is $C_1 N_i^{(1)} + C_2 N_i^{(2)}$ for any constants C_1 and C_2. For, by (a), $C_1 N_i^{(1)}$ and $C_2 N_i^{(2)}$ are solutions, and by (b), so is their sum.

Linear equations with constant coefficients

The simplest difference equation is a first-order linear equation with constant coefficients. Such an equation may be written

$$x_{n+1} = Rx_n, \quad n = 0, 1, 2, \ldots, \tag{10}$$

where R is a constant. We now call the unknown x_n instead of N_i, to accustom the reader to different notations. Note that

$$x_n = Rx_{n-1}, \quad n = 1, 2, 3, \ldots$$

is equivalent to (10).

Given the initial condition

$$x_0 = C, \quad C \text{ given}, \tag{11}$$

the successive terms in (10) may easily be found:

$$x_1 = RC, \ x_2 = R(RC) = R^2 C, \ x_3 = R^3 C, \ldots.$$

In general,

$$x_n = CR^n, \quad n = 0, 1, 2, \ldots. \tag{12}$$

For the second-order equation

$$x_{n+2} = Rx_{n+1} + Sx_n, \tag{13}$$

one can still find solutions proportional to a constant raised to a power, for if

$$x_n = k^n, \quad k \text{ a constant}, \tag{14}$$

then, on substitution into (13),

$$k^{n+2} = Rk^{n+1} + Sk^n \quad \text{or} \quad k^n(k^2 - Rk - S) = 0, \tag{15}$$

and k^n will be a solution for the two roots k_1 and k_2 of the quadratic equation

$$k^2 - Rk - S = 0. \tag{16}$$

By the superposition principle, not only are k_1^n and k_2^n solutions but also

$$x_n = C_1 k_1^n + C_2 k_2^n \tag{17}$$

is a solution, for arbitrary constants C_1 and C_2. In fact, it can be shown that (17) is the **general solution**, in that all solutions to (13) can be obtained by suitably specializing the constants C_1 and C_2. There is an exceptional case, when the quadratic (16) has two equal roots. For this case the reader is referred to texts such as that by Goldberg (1961).

Example 1: Find the general solution of

$$x_n - 5x_{n-1} + 6x_{n-2} = 0. \tag{18}$$

What solution satisfies the initial conditions $x_0 = 0$, $x_1 = 1$?

Solution: Assuming

$$x_n = k^n, \tag{19}$$

we find on substituting (19) into (18) that

$$0 = k^n - 5k^{n-1} + 6k^{n-2} = k^{n-2}(k^2 - 5k + 6).$$

Thus, k must satisfy

$$k^2 - 5k + 6 = 0, \quad \text{i.e.,} \quad (k-2)(k-3) = 0, \quad k = 2, 3.$$

The general solution is

$$C_1 2^n + C_2 3^n. \tag{20}$$

From the initial conditions,

$$0 = C_1 + C_2, \quad 1 = 2C_1 + 3C_2,$$

so that $C_1 = -1$, $C_2 = 1$, and the required solution is

$$x_n = 3^n - 2^n.$$

If the quadratic (16) has complex conjugate roots z_1 and z_2, then a more convenient form of the general solution is obtained by using de Moivre's theorem (A6.9) to write

$$z_1 = re^{i\theta}, \quad z_2 = re^{-i\theta},$$

$$z_1^n = r^n(\cos n\theta + i \sin n\theta), \quad z_2^n = r^n(\cos n\theta - i \sin n\theta).$$

It turns out that in this case the general solution can be written

$$x_n = r^n(C_1 \cos n\theta + C_2 \sin n\theta). \tag{21}$$

Example 2: Find the general solution of

$$x_n - 6x_{n-1} + 25x_{n-2} = 0. \tag{22}$$

Solution: If $x_n = k^n$, then k must satisfy

$$k^2 - 6k + 25 = 0, \quad k = \frac{6 \pm (36 - 100)^{1/2}}{2};$$

that is,

$$k = 3 + 4i, \quad 3 - 4i.$$

But [see (A6.11)]

$$3 + 4i = re^{i\theta} \quad \text{and} \quad 3 - 4i = re^{-i\theta} \quad \text{if } r = (3^2 + 4^2)^{1/2} = 5, \quad \theta = \tan^{-1}\tfrac{4}{3} \approx 0.93.$$

Thus, the general solution is (approximately)

$$x_n = 5^n[C_1 \cos(0.93n) + C_2 \sin(0.93n)]. \tag{23}$$

The general nth-order difference equation with constant coefficients

$$a_n x_n + a_{n-1} x_{n-1} + \ldots + a_0 x_0 = 0 \tag{24}$$

can also be solved by combining solutions of the simple form (14). Various values of k are obtained from the nth-order algebraic equation found by substitution of (14) into (24). Again, the case of multiple roots requires special attention.

Steady-state solutions and their stability

From now on we shall concentrate our discussion on the first-order equation $N_i = f(N_{i-1})$, although our ideas can apply to nth-order equations with little modification.

Of special interest are **steady-state** (or just "steady") solutions in which N_i keeps the same value "forever." Formally, a steady-state solution is such that

$$N_i = N \quad \text{for all } i, \quad N \text{ a constant.} \tag{25}$$

Such solutions satisfy the algebraic equation

$$N = f(N). \tag{26}$$

These solutions are particularly important, because if in successive years the population approaches closer and closer to a limit, then such a limit must be a steady-state solution. For if

$$\lim_{i \to \infty} N_i = N,$$

then (assuming that f is a continuous function)

$$\lim_{i \to \infty} N_i = \lim_{i \to \infty} f(N_{i-1}), \quad \text{i.e.,} \quad N = f\left(\lim_{i \to \infty} N_{i-1}\right) = f(N). \tag{27}$$

Given a steady-state solution, we often wish to determine its **stability**. To do this, we examine what happens if, in a given year, the population is near to the steady value in question. Analytically, we write

$$N_i = N + n_i, \tag{28}$$

so that the original equation $N_i = f(N_{i-1})$ implies

$$N + n_i = f(N + n_{i-1}). \tag{29}$$

We assume that n_{i-1} is small (it is then called a **perturbation**), so that indeed the population is close to its steady-state value N. The smallness of n_{i-1} makes reasonable the use of the two-term Taylor approximation

$$f(N + n_{i-1}) \approx f(N) + f'(N)n_{i-1}, \tag{30}$$

where

$$f'(N) = [df(x)/dx]_{x=N}. \tag{31}$$

Table A3.1. *Behavior of solutions to* $n_i = Rn_{i-1}$

	Values of R	Behavior of n_i/n_0
(i)	$R > 1$	Approaches infinity monotonically
	$R = 1$	Equals unity
(ii)	$0 < R < 1$	Approaches zero monotonically
	$R = 0$	Equals zero
(iii)	$-1 < R < 0$	Approaches zero in an oscillatory fashion
	$R = -1$	Oscillates between $+1$ and -1
(iv)	$R < -1$	Approaches infinity in an oscillatory fashion

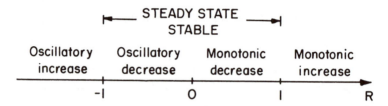

Figure A3.1. Qualitative behavior of perturbations n_i that are governed by $n_i = Rn_{i-1}$. (From Table A3.1, ignoring the special transitional cases in which $R = 1$ or $R = -1$.)

[See A2.7b).] Using (30), we can approximate (29) by

$$N + n_i = f(N) + f'(N)n_{i-1}. \qquad (32)$$

But by the steady-state equation (26), the N on the left side of (32) balances the $f(N)$ on the right. Thus, by limiting ourselves to the study of small perturbations, we obtain the following simplified version of the full equation (29) for n_i:

$$n_i = Rn_{i-1}, \quad R \equiv f'(N). \qquad (33)$$

Equation (33) is a first-order difference equation with constant coefficients whose general solution, as in (12), is $n_i = n_0 R^i$. The behavior of the solution as a function of R is given in Table A3.1. For $|R| < 1$, the perturbations approach zero as $i \to \infty$, so that the given steady-state solution is termed **stable**. For $|R| > 1$, perturbations increase in absolute value. Eventually the perturbations are predicted to become so large that the approximate equation (33) will no longer be valid, but the prediction of initial growth is trustworthy, so that the given steady-state solution is called **unstable**. See Figure A3.1.

A stability calculation for a second-order difference equation is outlined in Exercise 2.5.

Exercises

† 1. (a) If $x_{i+1} = -5x_i$, $x_0 = 1$, find x_3.

(b) If $x_{n+1} = \frac{1}{3}x_n$, $x_1 = 4$, find x_n.

(c) If $z_{k+1} = -\pi z_k$, $z_0 = 1$, find z_k.

2. The Fibonacci numbers are defined by

$$F_{n+1} = F_n + F_{n-1}, \quad n = 2, 3, \ldots; \quad F_1 = 1, F_2 = 1.$$

(a) Find $F_3, F_4, \ldots, F_9$.

† (b) Show that if the ratio F_{n+1}/F_n approaches a limit F, then $F = \frac{1}{2}(1 + \sqrt{5})$. [Fibonacci numbers and their generalizations arise in descriptive phyllotaxis (the study of the arrangement of leaves on a stem, florets on a pineapple, scales on a pine cone, etc.). Mitchison (1977) and Richter and Schranner (1978) have devised theoretical explanations for the observations.]

3. Prove by substitution that 2^n and 3^n are solutions of (18) but that 4^n is not.

† 4. (a) Find the general solution to $z_{n+2} - 3z_{n+1} + 2z_n = 0$.

(b) Find the solution to (a) that satisfies $z_1 = 0$, $z_2 = 2$.

† 5. (a) For what (complex) values of k is k^n a solution of

$$y_n - 4y_{n-1} + 13y_{n-2} = 0?$$

(b) Show that the general solution to the equation of (a) is

$$y_n = 13^{n/2}(C_1 \cos n\theta + C_2 \sin n\theta), \quad \text{where } \tan \theta = \frac{3}{2}.$$

† (c) For what values of C_1 and C_2 will the initial conditions $y_0 = 1$, $y_1 = \sqrt{13}$ be satisfied?

Appendix 4

Linear differential equations with constant coefficients[1]

First-order equations – constant coefficients

Most biologists are familiar with the simple differential equation

$$dx/dt = rx, \quad r \text{ a constant},\tag{1}$$

that describes the growth of a population under conditions where the number of births per individual per unit time, r, does not vary. The **general solution** to (1) is

$$x = x_0 \exp(rt), \quad x_0 \text{ a constant},\tag{2}$$

in the sense that any solution can be represented by suitably assigning the constant x_0. Verification that the solution is indeed general is beyond our scope here, but the fact that (2) is *a* solution is easily verified by direct substitution:

$$\frac{dx}{dt} = \frac{d}{dt}(x_0 e^{rt}) = x_0 \frac{d}{dt}(e^{rt}) = x_0 e^{rt} \frac{d}{dt}(rt) = rx_0 e^{rt} = rx.\tag{3}$$

Linear equations

A more complicated differential equation is

$$\alpha(t)\frac{d^2x}{dt^2} + \beta(t)\frac{dx}{dt} + \gamma(t)x = 0.\tag{4}$$

This equation is called **second-order**, because it contains the second-order derivative d^2x/dt^2, but no higher-order derivative. Equation (1) is a **first-order** differential equation; generally we can have an equation whose order is any positive integer n. Both (1) and (2) are **linear** differential equations, in that each term contains the unknown x or one of its derivatives, perhaps multiplied by a function of the independent variable

---

1 Readers who are unfamiliar with complex numbers (Appendix 6) may on first reading
  wish to take on faith that to the complex roots of (14) correspond the solutions (17).

t, but in no other combination. There are no terms such as x^8 or $(dx/dt)(d^2x/dt^2)$ or $\sin x$. A linear equation of nth order can be written

$$\alpha_n(t)\frac{d^n x}{dt^n}+\alpha_{n-1}\frac{d^{n-1}x}{dt^{n-1}}+\ldots+\alpha_1(t)\frac{dx}{dt}+\alpha_0(t)x=0. \tag{5}$$

If all the functions α_n are constants, then equation (5) is said to have **constant coefficients**.

For a linear equation, (a) *the sum of two solutions is also a solution,* and (b) *a constant multiple of a solution is also a solution.* We shall prove this for the second-order linear equation (4).

(a) Let the functions $x_1(t)$ and $x_2(t)$ be solutions of (4). Then

$$\alpha\frac{d^2x_1}{dt^2}+\beta\frac{dx_1}{dt}+\gamma x_1=0,\quad \alpha\frac{d^2x_2}{dt^2}+\beta\frac{dx_2}{dt}+\gamma x_2=0. \tag{6}$$

We wish to prove that $x_1(t)+x_2(t)$ is a solution; that is,

$$\alpha\frac{d^2(x_1+x_2)}{dt^2}+\beta\frac{d(x_1+x_2)}{dt}+\gamma(x_1+x_2)=0.$$

But using the basic properties of derivatives, the left side of the preceding equation equals

$$\alpha\frac{d^2x_1}{dt^2}+\alpha\frac{d^2x_2}{dt^2}+\beta\frac{dx_1}{dt}+\beta\frac{dx_2}{dt}+\gamma x_1+\gamma x_2=0,$$

where, for the last step, we have employed (6).

(b) If $x(t)$ satisfies (4), then for any *constant C,*

$$\alpha\frac{d^2(Cx)}{dt^2}+\beta\frac{d(Cx)}{dt}+\gamma(Cx)=\alpha C\frac{d^2x}{dt^2}+\beta C\frac{dx}{dt}+\gamma Cx$$

$$=C\left(\alpha\frac{d^2x}{dt^2}+\beta\frac{dx}{dt}+\gamma x\right)=0.$$

We can write the contents of the theorem just proved in a simple way if we denote the linear differential equation (5) by

$$L(x)=0,\quad \text{where } L(x)=\alpha_n(t)\frac{d^n x}{dt^n}+\ldots+\alpha_0(t)x. \tag{7}$$

According to our theorem,

$$L(x_1)=0\quad \text{and}\quad L(x_2)=0\quad \text{imply}$$

$$L(x_1+x_2)=0\quad \text{and}\quad L(Cx_1)=0, \tag{8}$$

where C is any constant. By using the theorem twice, we see that if $x_1(t)$ and $x_2(t)$ are solutions of $L(x)=0$, then for any constants C_1 and C_2,

$$L[C_1 x_1(t)]=0, \quad L[C_2 x_2(t)]=0, \quad \text{so } L[C_1 x_1(t)+C_2 x_2(t)]=0.$$

$C_1 x_1(t)+C_2 x_2(t)$ is called a **linear combination** of the functions x_1 and x_2. (Similarly, $5t+7t^2-11.6t^4$ is a linear combination of t, t^2, and t^4.) In terms of this definition, we can say that *for linear equations, new solutions can be formed by linear combinations of solutions already found.* This turns out to be the basis of the fact that the theory of linear differential equations is relatively easy.

Second-order linear equations – constant coefficients

We now discuss the second-order linear equation with constant coefficients

$$\frac{d^2x}{dt^2}+\beta\frac{dx}{dt}+\gamma x=0, \tag{9}$$

where β and γ are real (i.e., noncomplex) constants. Without loss of generality, we have assumed that the coefficient of d^2x/dt^2 is unity. This coefficient cannot be zero, because the equation is of second order. If the coefficient is α, $\alpha \neq 0$, we merely divide all coefficients by α to obtain the form (9).

It took a great mathematician (Euler) to come upon the right procedure for solving constant-coefficient differential equations, but it is easy for us to follow in his footsteps. As Euler suggested, we assume that (9) has solutions of the same exponential form that satisfied the simple first-order equation (1). If $y=\exp(mt)$, m a constant, is to satisfy (9), then

$$\frac{d^2(e^{mt})}{dt^2}+\beta\frac{d(e^{mt})}{dt}+\gamma e^{mt}=0, \quad \text{i.e., } m^2 e^{mt}+\beta m e^{mt}+\gamma e^{mt}=0. \tag{10}$$

Because $\exp(mt)$ is never zero, then for (10) to hold,

$$m^2+\beta m+\gamma=0. \tag{11}$$

Let us first consider the case $\beta^2-4\gamma>0$. Then there are two real roots of (11):

$$m_1=\tfrac{1}{2}[-\beta+(\beta^2-4\gamma)^{1/2}] \quad \text{and} \quad m_2=\tfrac{1}{2}[-\beta-(\beta^2-4\gamma)^{1/2}]. \tag{12}$$

There are two different exponential solutions: $\exp(m_1 t)$ and $\exp(m_2 t)$. It can be shown that the general solution can be written as a linear combination of these solutions:

$$x = C_1 e^{m_1 t} + C_2 e^{m_2 t}, \quad C_1 \text{ and } C_2 \text{ constants.} \tag{13}$$

When $\beta^2 - 4\gamma < 0$, the two roots take the form

$$m_1 = p + iq, \quad m_2 = p - iq, \tag{14}$$

where the real numbers p and q are given by

$$p \equiv -\tfrac{1}{2}\beta, \quad q \equiv \tfrac{1}{2}(4\gamma - \beta^2)^{1/2}. \tag{15}$$

The corresponding solutions are

$$x_1(t) \equiv e^{(p+iq)t} = e^{pt} e^{iqt} \quad \text{and} \quad x_2(t) \equiv e^{(p-iq)t} = e^{pt} e^{-iqt}.$$

By de Moivre's theorem (A6.9),

$$x_1(t) = e^{pt}(\cos qt + i \sin qt), \quad x_2(t) = e^{pt}(\cos qt - i \sin qt). \tag{16}$$

Because the equation under investigation, (9), is linear, the linear combinations $\tfrac{1}{2}x_1(t) + \tfrac{1}{2}x_2(t)$ and $(i/2)x_2(t) - (i/2)x_1(t)$ are also solutions. Those solutions are

$$e^{pt} \cos qt \quad \text{and} \quad e^{pt} \sin qt, \tag{17}$$

as can be verified by direct substitution (Exercise 3). It can be shown that the general solution can be written as a linear combination of these solutions:

$$x = C_1 e^{pt} \cos qt + C_2 e^{pt} \sin qt, \quad C_1 \text{ and } C_2 \text{ constants.} \tag{18}$$

To summarize, (13) and (12) give an explicit formula for the general solution of (9) when $\beta^2 > 4\gamma$, whereas (18) and (15) provide the general solution when $\beta^2 < 4\gamma$. When $\beta^2 = 4\gamma$, the two roots of the quadratic (11) are equal. See texts such as Boyce and DiPrima (1977) for the general solution in this exceptional case.

Although we shall not discuss it, the general solution to an nth-order equation with constant coefficients can be obtained by a straightforward generalization of what we have just done. In particular, we assume a solution of the form $\exp(mt)$ and find that m must be the solution of an nth-order polynomial equation that is analogous to the quadratic (11).

A system of two linear equations

Commonly met is the system of two first-order equations with constant coefficients

$$dx/dt = ax + by, \quad dy/dt = cx + dy; \quad a, b, c, d \text{ constants.} \tag{19a,b}$$

Here we seek the unknown functions $x(t)$ and $y(t)$. The method of solution that we shall use is to eliminate either $x(t)$ or $y(t)$, and thereby to arrive at a single linear equation with constant coefficients. For example, unless b is zero, we can solve (19a) for y, obtaining

$$y = b^{-1}\left(\frac{dx}{dt} - ax\right). \tag{20}$$

Substitution of (20) into (19b) yields

$$b^{-1}\left(\frac{d^2x}{dt^2} - a\frac{dx}{dt}\right) = cx + db^{-1}\left(\frac{dx}{dt} - ax\right) = 0,$$

or

$$\frac{d^2x}{dt^2} + \beta\frac{dx}{dt} + \gamma x = 0, \quad \text{where } \beta = -(a+d), \ \gamma = ad - bc. \tag{21a,b}$$

We have just provided formulas for obtaining the general solution of (21a). Once $x(t)$ has been determined, the required expression for $y(t)$ can at once be obtained from (20).

If $b = 0$ but $c \neq 0$, we can solve (19b) for x and substitute the result into (19a), obtaining

$$\frac{d^2y}{dt^2} + \beta\frac{dy}{dt} + \gamma y = 0, \quad \text{where } \beta = -(a+d), \ \gamma = ad,$$

an equation of exactly the same type as (21a). If $b = c = 0$, then (19) splits into two separate first-order equations,

$$dx/dt = ax, \quad dy/dt = dy,$$

that can at once be solved.

Final remark

For any particular example of the second-order linear equation with constant coefficients (9), we can, of course, substitute particular values into the formulas we have given, and thereby find the required solution. It is preferable not to have to rely on looking up specific formulas, but rather to adopt the principle of seeking exponential solutions. Similarly, for a particular example of a system like (19), it is better to apply the principle of eliminating one of the unknowns.

Exercises

† 1. (a) Find the general solutions to $dx/dt = 7x$ and $2(dy/dt) - 4y = 0$.
(b) Solve $dz/dt = -2z$, $z(0) = 2$.
(c) Find the general solutions to

$$\frac{d^2x}{dt^2} + 7\frac{dx}{dt} + 12x = 0 \quad \text{and} \quad 2\frac{d^2y}{dt^2} + 9\frac{dy}{dt} - 5y = 0.$$

(d) Solve $(d^2y/dt^2) + 2(dy/dt) - 3y = 0$. At $t = 0$: $y = 0$, $dy/dt = 2$.
(e) Solve $6(d^2z/dx^2) - 5(dz/dx) + z = 0$. At $x = 0$: $z = 1$, $dz/dx = 0$.

The next three problems involve oscillatory solutions.

† 2. For what values of the constant p are the solutions of the following equation oscillatory functions of the time t?

$$\frac{d^2y}{dt^2} + p\frac{dy}{dt} + y = 0.$$

What is the general solution in this situation?
3. Verify by substitution that the expressions of (17), given (15), satisfy (9).
4. Consider $d^2x/dt^2 - 4(dx/dt) + 13x = 0$.
(a) Show that the general solution is

$$x = e^{2t}(C_1 \cos 3t + C_2 \sin 3t).$$

(b) Verify by substitution that $e^{2t} \cos 3t$ is indeed a solution.
† (c) Find a particular solution that satisfies the initial conditions that at $t = 0$, $x = 3$ and $dx/dt = 12$.

Appendix 5

Phase-plane analysis

Trajectories in the phase plane

The purpose of this appendix is to describe some relatively easy methods for exploring the behavior of solutions to a general system of two ordinary differential equations for $x(t)$ and $y(t)$:

$$dx/dt = f(x,y), \quad dy/dt = g(x,y). \tag{1a,b}$$

It is helpful to regard the independent variable t as the time, though it may not have this meaning in any given application.

To illustrate ideas, we begin with the very simple example

$$dx/dt = x, \quad dy/dt = 2y, \tag{2a,b}$$

with the initial conditions

$$x(0) = x_0, \quad y(0) = y_0. \tag{3a,b}$$

The solution of the problem posed by (2) and (3) is

$$x = x_0 \exp(t), \quad y = y_0 \exp(2t). \tag{4a,b}$$

For any particular set of initial conditions, say $x_0 = 2$, $y_0 = 3$, we can visualize the solution by a pair of graphs, one for $x(t)$ and one for $y(t)$. See Figure A5.1. We can also display the results in another fashion by eliminating t from the solutions

$$x = 2e^t, \quad y = 3e^{2t}, \tag{5}$$

obtaining

$$y/3 = e^{2t} = (x/2)^2, \quad \text{i.e.,} \quad y = (3/4)x^2. \tag{6}$$

A graph of (6) will show the relationship between x and y (Figure A5.2). The way the solution develops with time can be indicated by placing heavy dots at selected times, as shown in Figure A5.2. *The direction that the point $x(t)$, $y(t)$ moves as time increases is indicated by arrowheads.*

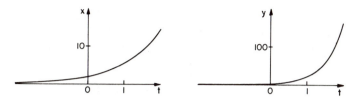

Figure A5.1. Graphs of the solutions $x(t)$ and $y(t)$ to the system (2) with initial conditions $x(0)=2$, $y(0)=3$.

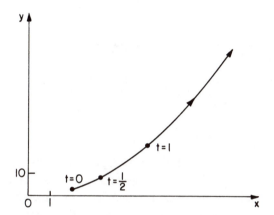

Figure A5.2. Another graph of the solutions to (2), showing the relationship between y and x. Successive heavy points on the graph correspond to times $t=0$, $t=\frac{1}{2}$, $t=1$ [i.e., from (5), $x=2$, $x=2e^{1/2}$, $x=2e$].

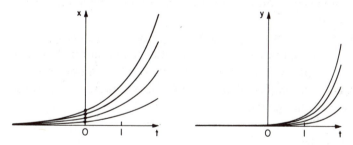

Figure A5.3. Graphs of $x=x_0 e^t$ for $x_0=1,2,3,4$, and of $y=y_0 e^{2t}$ for $y_0 = 1,2,3,4$.

It may be desirable to visualize solutions for several sets of initial conditions. One way to do this is to plot $x(t)$ and $y(t)$ for various initial conditions, as in Figure A5.3. Alternatively, we can plot the relationship between y and x implied by (4), namely,

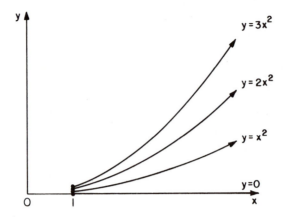

Figure A5.4. Several trajectories (7) of the system (2). As t increases, the state point $[x(t), y(t)]$ moves in the direction of the arrowheads, from the initial points $(1,0)$, $(1,1)$, $(1,2)$, $(1,3)$.

$$(y/y_0) = (x/x_0)^2 \quad \text{or} \quad y = (y_0/x_0^2)x^2, \tag{7}$$

for a number of different values of (x_0, y_0). Again, arrowheads are added to indicate the direction that (x, y) moves as t increases (Figure A5.4). Note that every point of the plane is interesected by a member of the family of curves (7); for any particular point (x_0, y_0) that we take, (7) gives the equation of the curve that passes through that point.

Often we wish to determine the behavior of all possible solutions to the system (1). Then we employ a plot like Figure A5.4 of the x–y plane or **phase plane**, indicating the behavior of all possible (x, y) curves (Figure A5.5).

As we have just mentioned for the particular case (2), any point in the phase plane can be regarded as an initial condition for (1). The solution curve that passes through this point is one of the **trajectories** of the system (1). The totality of all possible trajectories constitutes the **phase portrait** of the system.

Steady states

One of the most important classes of solutions to the system (1) is composed of the **steady states**

$$x(t) = \bar{x}, \quad y(t) = \bar{y}; \quad \bar{x}, \bar{y} \text{ constants.} \tag{8}$$

Because the derivative of a constant is zero, such solutions satisfy

$$f(\bar{x}, \bar{y}) = 0, \quad g(\bar{x}, \bar{y}) = 0. \tag{9}$$

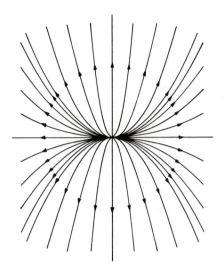

Figure A5.5. A phase portrait of system (2). Indicated on the graph are several trajectories (7), for positive and negative x_0 and y_0. (Of course, it is not feasible to draw *all* the trajectories.)

They represent situations in which the variables $x(t)$ and $y(t)$ can retain exactly the same values "forever."

The algebraic equations (9) may have several solutions. It is important to check all solutions for biological **admissibility**: Chemical concentrations must be nonnegative, probabilities must lie between zero and one, and so forth.

It is a fundamental fact about the phase plane that *trajectories can cross one another only at steady-state points*. To see this, note that the slope of a trajectory is given by

$$\frac{dy}{dx} = \frac{dy/dt}{dx/dt} = \frac{g(x,y)}{f(x,y)}. \tag{10}$$

At any point (x_0, y_0), then, the slope of any trajectory through the point will be given by the number

$$\frac{dy}{dx}\bigg|_{x=x_0, y=y_0} = \frac{g(x_0, y_0)}{f(x_0, y_0)}.$$

If trajectories crossed at (x_0, y_0), there would be more than one slope at that point, which is not consistent with the unique result just obtained. Some trouble might be anticipated if $f(x_0, y_0) = 0$, for then the slope is not defined. But as long as $g(x_0, y_0) \neq 0$,

$$\lim_{\substack{x \to x_0 \\ y \to y_0}} f(x,y) = 0 \quad \text{implies} \quad \lim_{\substack{x \to x_0 \\ y \to y_0}} \frac{g(x,y)}{f(x,y)} = \pm \infty. \tag{11}$$

This merely means that the trajectory at (x_0, y_0) is vertical. The only situation in which more than one result for the slope can be obtained on approaching (x_0, y_0) is when $g(x_0, y_0) = 0$ and $f(x_0, y_0) = 0$, but in that case (x_0, y_0) is a steady-state point.

As an example, note that dx/dt and dy/dt both vanish in (2) when and only when $x = 0$ and $y = 0$, and, indeed, trajectories in Figure A5.5 intersect at the origin and at no other point.

The simple arguments that we have employed do not rule out the possibility that trajectories might be *tangent* at a point that does not represent a steady state. But theorems concerning the uniqueness of solutions rule out this contingency, provided that the functions f and g of (1) satisfy certain weak conditions (Boyce and DiPrima, 1977, Section 7.1).

Phase portraits near steady states

The next step in determining phase-plane behavior is to look closely at the immediate neighborhood of the crucial steady-state points. This is done by introducing new variables $x'(t)$ and $y'(t)$, by

$$x'(t) = x(t) - \bar{x}, \quad y'(t) - y(t) - \bar{y}. \tag{12}$$

We obtain equations in terms of the new variables x' and y' by substituting into the original equations (1). This yields

$$\frac{dx'}{dt} = f(\bar{x} + x', \bar{y} + y'), \quad \frac{dy'}{dt} = g(\bar{x} + x', \bar{y} + y'). \tag{13}$$

Next comes the crucial step of approximating the equations by means of the Taylor formula (A2.8). Because x' and y' are assumed to be small disturbances (and hence are called **perturbations**), we retain only terms proportional to x' and y', and we neglect the relatively small higher-order terms proportional to $(x')^2$, $x'y'$, $(y')^2$ (the **second-order terms**), $(x')^3$, $(x'^2)y'$, and so forth. The result is

$$dx'/dt = ax' + by', \tag{14a}$$

$$dy'/dt = cx' + dy', \tag{14b}$$

where the constants a, b, c, and d are given by

$$a = \frac{\partial f(x,y)}{\partial x}\bigg|_{x=\bar{x},\, y=\bar{y}}, \quad b = \frac{\partial f(x,y)}{\partial y}\bigg|_{x=\bar{x},\, y=\bar{y}}$$

$$c = \frac{\partial g(x, y)}{\partial x} \bigg|_{x=\bar{x},\, y=\bar{y}}, \quad d = \frac{\partial g(x, y)}{\partial y} \bigg|_{x=\bar{x},\, y=\bar{y}}. \tag{15}$$

Determination of the behavior of solutions in the immediate neighborhood of a given steady-state point $(\bar{x}, \bar{y})$ has been reduced to the study of the system (14) for the perturbations $x'(t)$ and $y'(t)$. Appendix 4 gives expressions for the solutions to (14) in terms of two arbitrary constants. These constants can be selected by the initial conditions $x'(0) = x_0$, $y'(0) = y_0$, so that the perturbation begins at any desired point (x_0, y_0) near $(\bar{x}, \bar{y})$. If perturbations decay (i.e., if the perturbation decreases toward zero) no matter what initial point we choose, then the steady-state point is called **stable**. We shall also meet cases in which almost all perturbations that start near $(\bar{x}, \bar{y})$ tend to increase in amplitude. In this case the steady-state point is termed **unstable**. The idea behind the definition is that in practice a given system is perturbed over and over again by outside disturbances, so that eventually a growing perturbation will surely be encountered.

Not only the stability or instability of a given steady-state point is of interest, but also the general form of the solution in the neighborhood of this point. There are several different cases, as we shall now show by examining the various possible qualitative behaviors of the solutions to (14).

It is shown in Appendix 4 that the solutions to the system (14) depend on the four coefficients a, b, c, and d only through the two parameters

$$\beta = -(a+d) \quad \text{and} \quad \gamma = ad - bc. \tag{16}$$

In citing the various solutions we shall assume that $b \neq 0$. It turns out that our conclusions also hold if $b = 0$ (Exercise 4).

If $\beta^2 > 4\gamma$, then by (A4.13), (A4.20), and (A4.12), the solutions to (14) are

$$x'(t) = C_1 e^{m_1 t} + C_2 e^{m_2 t}, \tag{17a}$$

$$y'(t) = b^{-1}(m_1 - a)C_1 e^{m_1 t} + b^{-1}(m_2 - a)C_2 e^{m_2 t}. \tag{17b}$$

Here the (real) numbers m_1 and m_2 are given by

$$m_1 = \tfrac{1}{2}[-\beta + (\beta^2 - 4\gamma)^{1/2}] \quad \text{and} \quad m_2 = \tfrac{1}{2}[-\beta - (\beta^2 - 4\gamma)^{1/2}], \tag{17c}$$

and C_1 and C_2 are arbitrary constants. The behavior of the solutions depends strongly on the signs of m_1 and m_2.

In analyzing (17), we suppose, first, that $\gamma > 0$ and $\beta < 0$. Then both m_1 and m_2 are positive, and the solutions (17) (for all C_1 and C_2) are

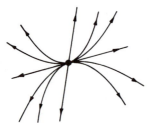

Figure A5.6. Typical behavior near an unstable node. In a stable node, the arrows, which denote the direction of motion of $[x(t), y(t)]$, are reversed. Only a few sample trajectories are depicted.

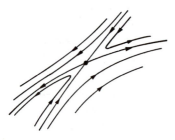

Figure A5.7. Typical behavior near a saddle point.

composed of functions that grow exponentially in time. In this case the steady-state point is unstable. The trajectories near the unsteady-state point will be similar to those in Figure A5.5, for here, too, both variables grow exponentially in time. An example of how the trajectories appear in general for such an **unstable node** appears in Figure A5.6. The qualitative features are the same as in the particular example depicted in Figure A5.5, but Figure A5.5 is more symmetric.

If $\gamma > 0$ but $\beta > 0$, then $m_1 < 0$ and $m_2 < 0$. Now both exponentials in (17a) decay in time, and the steady-state point is stable. The corresponding phase portrait for this **stable node** is qualitatively identical with that of the previous case, except that all trajectories head toward the critical point.

If $\gamma < 0$, then one of the two exponents is positive and one negative, whatever the sign of β. For definiteness, suppose that $\beta > 0$. Then $m_1 > 0$ and $m_2 < 0$. If $C_1 = 0$ in (17a), then the solutions tend to zero as $t \to \infty$. Generally, $C_1 \neq 0$, and solutions tend to infinity as $t \to \infty$. Figure A5.7 shows the appearance of some trajectories in this case of a **saddle point**. A particular example that allows easy calculation of the phase portrait is given in Exercise 2.

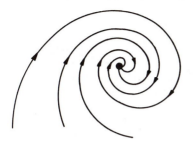

Figure A5.8. Typical behavior near a stable spiral (or focus). In an unstable spiral, the arrows are reversed. Spirals may be either clockwise (as shown) or counterclockwise.

Let us now turn to the cases in which $\beta^2 < 4\gamma$. Now, according to (A4.18) and (A4.20), the solutions to (14) are given by

$$x'(t) = C_1 e^{pt} \cos(qt) + C_2 e^{pt} \sin(qt), \tag{18a}$$

$$y'(t) = b^{-1}(pC_1 + qC_2 - aC_1)e^{pt} \cos(qt) + b^{-1}(pC_2 - qC_1 - aC_2)$$

$$\times e^{pt} \sin(qt), \tag{18b}$$

where [by (A4.15)]

$$p = -\beta/2, \quad q = (4\gamma - \beta^2)^{1/2}/2. \tag{18c}$$

The presence of the terms $\sin(qt)$ and $\cos(qt)$ gives the solutions an oscillatory character. The factor $\exp(pt) \equiv \exp(-\beta/2)t$ induces an overall growth (instability) or decay (stability), depending on whether β is negative or positive. Typical **spiral** trajectories are shown in Figure A5.8. Also see Exercise 3.

To summarize the nomenclature, when both m_1 and m_2 of (17) have the same sign, the steady-state point is called a **node**; when they have opposite signs, the term **saddle point** is employed. When complex exponents occur, and (18) results, then the steady-state point is called a **spiral point** or a **focus**. The adjectives *stable* and *unstable* are added as appropriate.

We stress that the phase portraits of the trajectories that are described by (17) or (18) (Figures A5.5–A5.8) are relevant to the original problem only in the close neighborhood of the steady-state point in question. The reason is that the crucial linearization step in our analysis limited us to a discussion of small perturbations from the steady state.

We have not treated the case $\beta = 0$, $\gamma > 0$. Here the solutions of (14) are purely oscillatory, but effects are in such delicate balance that the behavior of solutions to the unperturbed problem (13) might well be altered even by the very small nonlinear terms that we have omitted in

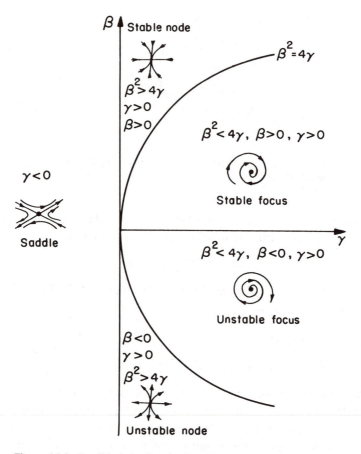

Figure A5.9. Possible behaviors in the neighborhood of steady-state points, found from analyzing perturbation equations $dx'/dt = ax' + by'$, $dy'/dt = cx' + dy'$; $\beta \equiv -(a+d)$, $\gamma \equiv ad - bc$.

our analysis. Thus, for our purposes, it is best to regard $\beta = 0$ as a transition line and not concern ourselves with the phase portrait in this situation. Similarly, we shall not discuss the transition cases $\gamma = 0$ and $\beta = 4\gamma^2$.

All our results concerning the different qualitative behaviors near steady points are summed up in Figure A5.9. It is now straightforward to determine the behavior of trajectories in the neighborhood of any steady-state solution $(\bar{x}, \bar{y})$ of the original equations (1). We calculate the coefficients a, b, c, and d by (15) or any other linearization procedure, calculate β and γ according to (16), and then refer to Figure A5.9 for the final answer.

Note that the steady states are stable if $\beta > 0$ and $\gamma > 0$ and are unstable *if either* $\beta < 0$ *or* $\gamma < 0$.[1] This result can also be interpreted as a *condition on the coefficients in* (14) *that give decaying or growing solutions.*

Example: Find the steady states possessed by the following equations for $v(t)$ and $s(t)$:

$$\frac{dv}{dt} = v(\delta e^{-s} - \rho), \quad \frac{ds}{dt} = \frac{\theta v}{1+v} - s; \quad v \geqslant 0, s \geqslant 0. \tag{19a,b}$$

Assume that the positive constants δ, ρ, and θ satisfy

$$\theta > \ln(\delta/\rho), \quad \delta > \rho. \tag{20a,b}$$

Examine the behavior in the neighborhood of the steady states by analyzing the perturbation equations. In particular, determine whether the steady states are stable or unstable.

Remark: The foregoing equations are taken from the biomathematics literature. Their significance for certain phenomena connected with liver regeneration is explored in Exercise 13, but the biological meaning of the equations will not be taken into account here.

Solution: The steady-state values of v and s are found from the equations

$$v(\delta e^{-s} - \rho) = 0, \quad \frac{\theta v}{1+v} - s = 0. \tag{21a,b}$$

One solution of (21a) is $v = 0$, and if $v = 0$, then (21b) implies that $s = 0$. If $v \neq 0$, a steady-state solution $(\bar{v}, \bar{s})$ is given by

$$\delta e^{-\bar{s}} = \rho, \quad \frac{\theta \bar{v}}{1+\bar{v}} = \bar{s}. \tag{22a,b}$$

Solving (22a) for $\bar{s}$ and (22b) for $\bar{v}$ in terms of $\bar{s}$, we obtain the following expressions for the steady-state solution $(\bar{v}, \bar{s})$:

$$\bar{s} = \ln(\delta/\rho), \quad \bar{v} = \bar{s}/(\theta - \bar{s}). \tag{23a,b}$$

[Note that (20a) ensures the positivity of $\bar{v}$.] In summary, there are two steady-state solutions

$$v = 0, \quad s = 0 \quad \text{and} \quad v = \bar{v}, \quad s = \bar{s}. \tag{24a,b}$$

Adapting (14) and (15) to our purposes, we see that the stability of the steady states is governed by equations of the form

$$\frac{dv'}{dt} = av' + bs', \quad \frac{ds'}{dt} = cv' + ds'. \tag{25a,b}$$

The constants a, b, c, and d are determined by

$$a = \frac{\partial f}{\partial v}, \quad b = \frac{\partial f}{\partial s}, \quad c = \frac{\partial g}{\partial v}, \quad d = \frac{\partial g}{\partial s}, \tag{26}$$

where

$$f(v,s) = v(\delta e^{-s} - \rho), \quad g(v,s) = \frac{\theta v}{1+v} - s, \tag{27a,b}$$

---

1 Those who have studied matrices will recognize $-\beta$ and γ as the trace and determinant of the coefficient matrix (15). Solutions to (14) decay with time, provided the eigenvalues of this matrix have negative real parts.

and where the derivatives must be evaluated at the appropriate steady-state values of v and s. We find in general that

$$\frac{\partial f}{\partial v} = \delta e^{-s} - \rho, \quad \frac{\partial f}{\partial s} = -\delta v e^{-s}, \tag{28a,b}$$

$$\frac{\partial g}{\partial v} = \theta(1+v)^{-2}, \quad \frac{\partial g}{\partial s} = -1. \tag{28c,d}$$

For the "trivial" steady-state solution (24a), we must evaluate the derivatives at $v = 0$, $s = 0$. This gives

$$a = \delta - \rho, \quad b = 0, \quad c = \theta, \quad d = -1.$$

The corresponding version of the linearized equations (25) is

$$\frac{dv'}{dt} = (\delta - \rho)v', \quad \frac{ds'}{dt} = \theta v' - s'. \tag{29}$$

Referring to the definitions (16), we find

$$\beta \equiv -(a+d) = -\delta + \rho + 1, \quad \gamma \equiv ad - bc = -\delta + \rho.$$

Because γ is negative, by (20b), there is no need for further calculations. The steady-state point $(0,0)$ is a saddle point (see Figure A5.9), and in particular it is unstable.

For the "nontrivial" steady-state point (24b), the derivatives must be evaluated when s and v take the constant values $\bar{s}$ and $\bar{v}$ given by (23). It is often convenient to keep in mind the original relationships between the constants given by (22), and this is the case here. Using (22a) in the evaluation of the constants a and b, we find that

$$a = 0, \quad b = -\rho\bar{v}, \quad c = \theta(1+\bar{v})^{-2}, \quad d = -1.$$

Now

$$\beta = 1, \quad \gamma = \rho\bar{v}\theta(1+\bar{v})^{-2}.$$

Because $\beta > 0$ and $\gamma > 0$, this steady-state point is certainly stable (Figure A5.9). It is a stable node if $\beta^2 > 4\gamma$, that is, if

$$\theta < (1+\bar{v})^2 / (4\rho\bar{v}), \tag{30}$$

and a stable focus if the preceding inequality is reversed.

Qualitative behavior of the phase plane

Experience indicates that the best way to obtain an understanding of the qualitative behavior of the solutions to a given pair of equations of the form (1) is to sketch trajectories in the phase plane. The following procedure is recommended.

I. Find the (admissible) steady-state points, and examine their stability. (Are they nodes, saddles, etc.?)

II. Plot the **horizontal nullclines**, the curve or curves where $g(x, y) = 0$. Indicate by small line segments that trajectories have horizontal tangents along these curves [because $dy/dx = g/f = 0$; see (10)].

III. Plot the **vertical nullcline** curve, or curves, where $f(x, y) = 0$, and indicate that trajectories have vertical tangents along these curves. Indicate by heavy dots each place where $f = 0$ intersects $g = 0$; these are steady-state points.

IV. Indicate by arrows whether the horizontal tangents point to the left $(dx/dt < 0)$ or to the right $(dx/dt > 0)$ and whether the vertical tangents point upward $(dy/dt > 0)$ or downward $(dy/dt < 0)$. To do this, use the fact that dx/dt and dy/dt will "usually" change sign when the nullclines are crossed, as described later.

V. Place further small arrows to indicate the direction of trajectories along the axes $x = 0$ and $y = 0$.

VI. Try to combine all the foregoing information into a consistent picture, remembering that trajectories can intersect only at steady-state points, and keeping in mind the required qualitative behavior (from I) near these points. (If possible, it is very helpful to sketch several particular trajectories with the aid of a computer.)

Example: Sketch several representative phase-plane trajectories for the equations (19) subject to the conditions (20).

Solution: In our case, the x-y plane becomes the v-s plane. We have already carried out step I. Turning to step II, horizontal tangents $(ds/dt = 0)$ occur along the curve $s = \theta v/(1 + v)$, which is plotted in Figure A5.10. Vertical tangents occur when

$$v(\delta e^{-s} - \rho) = 0, \quad \text{i.e.,} \quad v = 0 \quad \text{or} \quad s = \ln(\delta/\rho),$$

and this, too, is indicated on Figure A5.10. The steady-state points are indicated by heavy dots.

The nullclines generally perform the useful function of delineating regions where (in this case) ds/dt and dv/dt have constant sign. This property follows from a fact used in Chapter 3: A function changes sign only when it passes through zero or infinity, or (rarely) when it jumps discontinuously between positive and negative values.

In our case, the functions giving dv/dt and ds/dt in (19) are continuous. The right side of (19b) becomes infinite at $v = -1$, but this is outside the region $v \geqslant 0$, $s \geqslant 0$ in which we are interested. Thus, ds/dt, for example, can change sign only on crossing the horizontal nullcline $s = \theta v/(1 + v)$ that is depicted in Figure A5.10. From (19b) we see that ds/dt is positive when $s = 0$ and negative when s is sufficiently large. Thus, ds/dt is in fact negative for all points above the nullcline and positive for all points below it. (This result also follows from the observation that ds/dt is an increasing function of v.)

Because $\delta > \rho$, (19a) shows that dv/dt is positive when $s = 0$; dv/dt is negative when s is large. We conclude that dv/dt negative above the nullcline $s = \ln(\delta/\rho)$ and positive below it, as indicated in Figure A5.10. (How the signs of dv/dt are

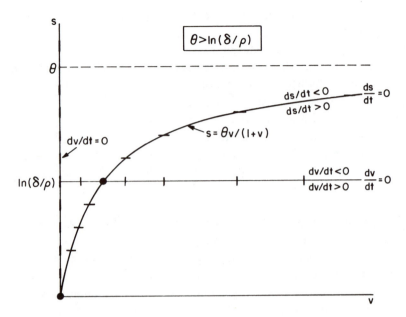

Figure A5.10. Horizontal ($ds/dt=0$) and vertical ($dv/dt=0$) nullclines for equations (19) subject to condition (20). Steady-state points at the intersections of nullclines are marked with heavy dots.

related to the nullcline can also be determined at once from the fact that when $v > 0$, dv/dt is a decreasing function of s.)

With the information just obtained, we can easily carry out steps IV and V, indicating the directions of various key trajectories. For example, as shown in Figure A5.11, the vertical trajectories on the right portion of $s = \ln(\delta/\rho)$ point upward – for they are in a region [below $s = \theta v/(1+v)$] where $ds/dt > 0$. To give another example, all points on the v axis are in a region where $ds/dt > 0$ and $dv/dt > 0$. At this axis, trajectories consequently point upward to the right (s increasing and v increasing), as shown in Figure A5.11.

We are now ready for the final step VI of incorporating our findings into a plausible sketch of the phase plane. All the information we have obtained has been depicted in Figure A5.11, except for our knowledge of behavior near the steady-state points. We know that $(0,0)$ is a saddle, and we observe in Figure A5.11 the characteristic behavior of a special trajectory heading inward, but with the general trajectory heading away from this steady-state point. The nontrivial steady-state point $(\bar{v}, \bar{s})$ is stable, so that trajectories must head inward. From (30), these trajectories must spiral inward when they are close to $(\bar{v}, \bar{s})$ if θ is large, and they must head directly inward if θ is small. Consistent with all our information are the computer plots of Figure A5.12, which show all trajectories heading toward $(\bar{v}, \bar{s})$ – with a pronounced spiraling character if θ is large.

Exercises

> 1. Show that the directions of the arrowheads on the trajectories in Figure A5.5 correctly give the change as time increases of $x(t)$ and

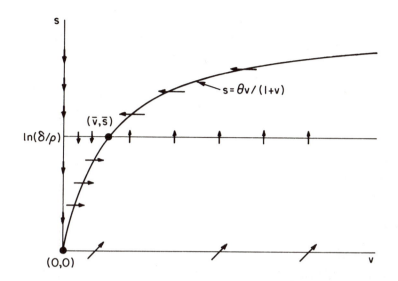

Figure A5.11. Continuation of the phase-plane analysis started in Figure A5.10. The directions taken by solutions at points of horizontality and verticality are marked, as is the direction of a solution crossing the positive v-axis. These directions are determined from information in Figure A5.10 on regions where dv/dt and ds/dt are positive and negative.

$y(t)$ according to (4). Check initial conditions (x_0, y_0) corresponding to all four quadrants.

2. (a) Solve for $x(t)$ and $y(t)$:

$$\frac{dx}{dt} = x, \quad \frac{dy}{dt} = -y; \quad x(0) = x_0, \quad y(0) = y_0. \tag{31}$$

(b) Show that

$$xy = x_0 y_0.$$

(c) Plot several examples of trajectories in the x–y plane with arrowheads indicating how $x(t)$ and $y(t)$ change as t increases.

3. (a) Show that the general solution of

$$\frac{dx}{dt} = x - y, \quad \frac{dy}{dt} = x + y, \tag{32}$$

is

$$x = e^t(C_1 \cos t + C_2 \sin t), \quad y = e^t(C_1 \sin t - C_2 \cos t).$$

(b) [Familiarity with polar coordinates is required.] Introduce new arbitrary parameters A and α by

$$C_1 = A \cos \alpha, \quad C_2 = A \sin \alpha.$$

Show that

$$x = Ae^t \cos(t - \alpha), \quad y = Ae^t \sin(t - \alpha),$$

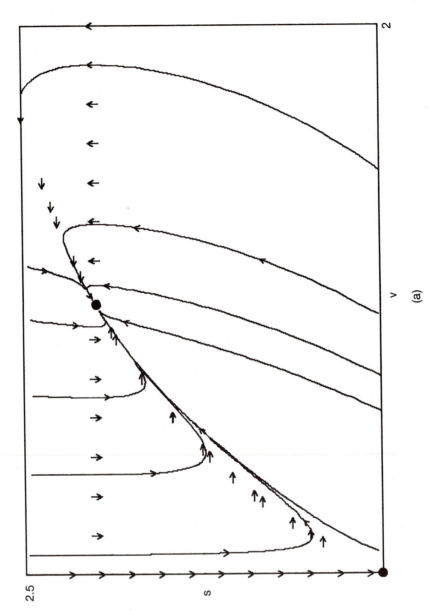

Figure A5.12. Computer-generated trajectories that supplement the phase-plane analysis started in Figure A5.10, where the nontrivial steady-state point is (a) a stable node or (b) a stable focus. In (a), $\rho = 0.1$, $\delta = 0.73$, $\theta = 4$. In (b), $\rho = 10$, $\delta = 73$, $\theta = 4$. This figure was drawn by G. Odell with his automatic phase portrait generator.

(a)

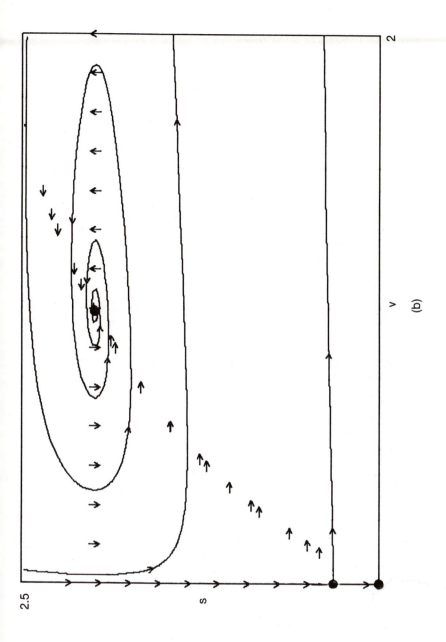

(b)

so that in polar coordinates (r, θ),

$r = Ae^t$, $\quad \theta = t - \alpha$.

(c) Sketch examples of the solution in (b) for $\alpha = 0$, $A = 1, 2, 3$.

4. Give counterparts of (17a–c) when $b = 0$, $c \neq 0$. What if $b = c = 0$?

† 5. Suppose that a steady state of (1) is stable, so that $\beta > 0$, $\gamma > 0$. Show that if instability ensues because $\beta(\gamma)$ changes sign, then the instability is of oscillatory (monotonic) nature.

6. (a) Show that (30) is equivalent to $\theta < 4\rho \bar{s}^2/(4\rho \bar{s} + 1)$.

(b) Consider the equations $dx/dt = f(x, y)$, $dy/dt = g(x, y)$, for $x \geqslant 0$, $y \geqslant 0$. Suppose that $(0, 0)$ is a stable node and that the only other steady-state point is at $(1, 1)$. Sketch a few trajectories in the phase plane in the two cases in which $(1, 1)$ is (i) an unstable node or (ii) an unstable focus.

† 7. Two chemicals have concentrations $C_1(t)$ and $C_2(t)$. Their reaction is described by equations of the form

$$dC_1/dt = f(C_1, C_2), \quad dC_2/dt = g(C_1, C_2).$$

Typical graphs of (C_1, C_2) as a function of t are given in Figure A5.13. Steady-state points are marked by heavy dots. Which of these is stable, and which unstable? Describe the qualitative behavior of the solutions.

8. Consider $dA/dt = aA - a_1 A^3$; $a > 0$, $a_1 > 0$. Show that $A = \sqrt{(a/a_1)}$ is a stable steady state.

9. Under certain circumstances the interaction of two species of micro-organisms can be described by differential equations of the form

$$dn_1/dt = f(n_1, n_2), \quad dn_2/dt = g(n_1, n_2),$$

where $n_1(t)$ and $n_2(t)$ are the populations at time t. It turns out that the equations have four steady states (A, B, C, D), as shown by heavy dots in Figure A5.14. The behaviors of trajectories near the steady states and along the axes are as shown.

(a) Which of the steady states is stable, and which unstable?

(b) What is the expected form of $n_1(t)$ and $n_2(t)$ near B? [Sample answer, correct in general idea, wrong in detail: "Near B, $n_1(t) \approx \alpha t^{-\beta}$, $n_2(t) \approx \gamma t^{-\delta}$; α, β, γ, and δ constants; $\beta > 0$, $\delta > 0$."]

(c) Sketch in more trajectories, so that the qualitative behavior of n_1 and n_2 becomes clear. Describe this behavior in a sentence or two.

10. Consider the following model for interactions between a predator (density y) and its prey (density x):

$$\frac{dx}{dt} = ax - bx^2 - cxy, \quad \frac{dy}{dt} = -ey + hxy. \tag{33}$$

Here, a, b, c, e, and h are positive parameters.

† (a) Give possible biological meaning to these equations.

(b) Show that the possible steady states are

(i) $x = 0$, $y = 0$; (ii) $x = \dfrac{e}{h}$, $y = \dfrac{a}{c} - \dfrac{be}{ch}$; (iii) $x = \dfrac{a}{b}$, $y = 0$.

Are these three steady states always biologically significant?

(c) Show that $(0, 0)$ is a saddle point.

(d) If $\bar{x} = x - x^*$ and $\bar{y} = y - y^*$ are perturbations from a steady state (x^*, y^*), show that the equations linearized about this steady state are

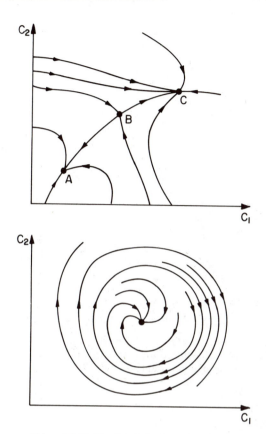

Figure A5.13. Possible behaviors of trajectories in the phase plane – for Exercise 7.

$$\frac{d\tilde{x}}{dt} = (a - 2bx^* - cy^*)\tilde{x} - (cx^*)\tilde{y},$$

$$\frac{d\tilde{y}}{db} = (-e + hx^*)\tilde{y} + (cy^*)\tilde{x}.$$

(e) For the steady state (ii), show that the linearized equations are

$$\frac{d\tilde{x}}{dt} = \left(-\frac{be}{h}\right)\tilde{x} - \left(\frac{ce}{h}\right)\tilde{y}, \quad \frac{d\tilde{y}}{dt} = \left(a - \frac{be}{h}\right)\tilde{x}.$$

For the case $a = 4$, $b = 2$, $c = h = e = 1$, show that this steady state is a stable spiral.

11. Models for cell growth often assume that a cell can be in various states, with switching (transfer) between one state and another. The present model assumes that there are two states, in only one of which is there proliferation (cell division). If $P(t)$ and $Q(t)$ represent the concentrations of cells in the two states, the following equation. can be assumed,

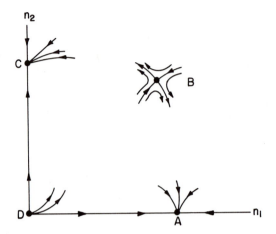

Figure A5.14. Partial sketch of phase-plane behavior for Exercise 9.

where all the Greek letters represent positive constants:

$$dP/dt = (\gamma - \delta P)P - \alpha P + \beta Q, \quad dQ/dt = \alpha P - (\lambda + \beta)Q. \tag{34a,b}$$

(a) Which is the proliferating state, P or Q? Why do you think so? Give a brief description of what is assumed in the model.
(b) Show that if the dimensionless variables

$$p = \frac{P}{\gamma/\delta}, \quad q = \frac{Q}{\alpha\gamma/[\delta(\lambda+\beta)]}, \quad \tau = \alpha t,$$

are introduced, the equations of (34) become

$$dp/d\tau = Gp(1-p) - p + Hq, \quad dq/d\tau = F(p-q), \tag{35a,b}$$

where

$$F = \frac{\lambda+\beta}{\alpha}, \quad G = \frac{\gamma}{\alpha}, \quad H = \frac{\beta}{\lambda+\beta}.$$

(c) Find all possible biologically meaningful steady states $(\bar{p}, \bar{q})$ of (35a) and (35b).
(d) Show that small perturbations p' and q' from a steady state $(\bar{p}, \bar{q})$ satisfy

$$dp'/d\tau = [G(1-2\bar{p}) - 1]p' + Hq',$$

$$dq'/d\tau = \qquad Fp' - Fq'.$$

(e) Discuss the case $1 - G < H$: Analyze the stability of the steady-state points, find the isoclines, draw little arrows indicating local behavior of trajectories on the isoclines and the axes, sketch several trajectories, conjecture qualitative behavior.
(f) Repeat (e) when $1 - G > H$.
(g) What general conclusions can be drawn from the analysis? [For further reading about this type of model, see Eisen and Schiller (1977).]

12. This problem concerns a model by McCarley and Hobson (1975) that is intended to explain regular oscillations in cat neuron activity that are associated with the sleep cycle. Activity is measured in number of electrical discharges per second. Two groups of cells are considered. If $x(t)$ and $y(t)$ are the activity levels of the two cell groups, the authors propose the equations

$$dx/dt = ax - bxy, \quad dy/dt = -cy + dxy, \qquad (36a,b)$$

where a, b, c, and d are positive constants.

(a) The following are excerpts from the paper that explain why the particular equations were chosen. A: "Evidence that the rate of change of activity levels...is proportional to the current level of activity." B: "Nonlinear interaction was to be expected. We model this effect... in accord with the reasonable physiological postulate that the effect of an excitatory or inhibitory input to the two populations will be proportional to the current level of discharge activity." Which terms are justified by part A and which by part B? Which are the excitatory influences, and which the inhibitory?

(b) Show that equations (36) have a steady-state solution in which neither x nor y is zero. Show further that if $m(t)$ and $n(t)$ are the departures of $x(t)$ and $y(t)$ from steady-state values, then linearized equations for m and n have the form

$$dm/dt = -\alpha n, \quad dn/dt = \beta m. \qquad (37)$$

Express α and β in terms of the original constants a, b, c, and d.

(c) Find the general solution to (37). The solution is oscillatory. Find the period of oscillation in terms of α and β.

(d) Complete an analysis of steady states and their stability. Sketch a few trajectories in the x–y (phase) plane.

13. This problem is concerned with a theory of liver regeneration due to Bard (1978, 1979). Normally the rate of cell division in the liver is very low, but if (in the rat) up to two-thirds of the liver is removed, then the liver grows back to its original size in about a week. Bard discusses two theories. One theory, based on the assumed existence of a growth stimulator, predicts that the liver volume V will overshoot its normal value before finally settling down to a steady state. Such an overshoot has not been observed. Here we shall show something of how an alternative inhibitor model can account for the facts.

Bard assumes that liver cells are prevented from dividing by an inhibitor of short half-life. The inhibitor is synthesized by the liver at a rate proportional to its size and is secreted into the blood, where its concentration is the same as in the liver. Let $V(t)$ be the volume of the liver and $s(t)$ the concentration of the inhibitor. Bard postulates the equations

$$\frac{dV}{dt} = V[f(S) - r], \quad \frac{dS}{dt} = \frac{pV}{W+V} - qS. \qquad (38a,b)$$

W is a constant, the blood volume; r, p, and q are also constants.

(a) Here are some questions about the meaning of the equations.

(i) The function f is assumed to have a negative derivative. Why?

(ii) In terms of the model, what is the mathematical translation of "short half-life"?

(iii) Explain carefully why $V + W$ appears where it does in (38b).

(b) Given that unique positive constants $\bar{V}$ and $\bar{S}$ are defined by

$$f(\bar{S})=r, \quad p\bar{V}/(\bar{V}+W)=q\bar{S},$$

define V' and S' by

$$V(t)=\bar{V}+V'(t), \quad S(t)=\bar{S}+S'(t).$$

Substitute into (38), and show that the linearized equations for V' and S' have the form

$$dV'/dt=-\gamma S', \quad dS'/dt=\alpha V'-qS', \tag{39a,b}$$

for certain positive constants γ and α.

(c) Bard claims that if q is sufficiently large, then the liver volume will return to its original value after some of it is removed, without oscillation. What support can you give to this claim, starting from equations (39a) and (39b)?

(d) Remembering that the derivative of f is negative, show by means of a sketch that a unique value of $\bar{S}$ can be determined from $f(\bar{S})=r$ provided that $f(0)>r$ and that otherwise there is no solution. Show that the corresponding value of $\bar{V}$ is "unbiological" unless $p>qS$.

(e) Find the second possible steady state.

(f) The second steady state is stable if $f(0)<r$ and unstable if $f(0)>r$. From this information and your conclusions from (d), without further calculation, make reasonable guesses about the phase-plane behavior, and hence the qualitative behavior, in the two cases $f(0)<r$ and $f(0)>r$, $p>q\bar{S}$. Is the behavior biologically reasonable?

(g) What happens when $f(0)>r$, $p<q\bar{S}$?

(h) Consider the particular case $f(S)=Ae^{-aS}$. Show that introduction of appropriate dimensionless variables reduces (38) to (19). How does the text's analysis of (19) relate to the foregoing analysis of (38)?

14. This problem concerns the following equations for $A(t)$ and $B(t)$:

$$dA/dt=aA-a_1A^3-a_2AB^2, \quad dB/dt=bB-b_1BA^2-b_2B^3.$$

All the coefficients are assumed to be positive. We shall study the equations only for nonnegative values of $A(t)$ and $B(t)$. The results are employed in the supplement to Chapter 8.

(a) Discuss the suitability of using these equations as a model for the interaction of two populations of bacteria.

(b) Show that there are four possible steady-state solutions:

   I: $A=B=0$;

   II: $A=0, \quad B^2=b/b_2$;

   III: $B=0, \quad A^2=a/a_1$;

   IV: $A^2=(a_2b-ab_2)/(a_2b_1-a_1b_2)\equiv\xi^2$,

        $B^2=(ab_1-a_1b)/(a_2b_1-a_1b_2)\equiv\eta^2$.

(c) Denote departures from the ith steady state by $A_i(t)$, $B_i(t)$. Show that the linearized perturbation equations are the following (where $\dot{}\equiv d/dt$):

   I: $\dot{A}_1=aA_1, \quad \dot{B}_1=bB_1$;

   II: $\dot{A}_2=b_2^{-1}(ab_2-a_2b)A_2, \quad \dot{B}_2=-2bB_2$;

III: $\dot{A}_3 = -2aA_3, \quad \dot{B}_3 = -a_1^{-1}(ab_1 - a_1 b)B_3;$

IV: $\dot{A}_4 = -2a_1 \xi^2 A_4 - 2a_2 \xi \eta B_4, \quad \dot{B}_4 = -2b_1 \xi \eta A_4 - 2b_2 \eta^2 B_4.$

(d) Examine the stability of the steady states, and demonstrate that there are four different qualitative possibilities:

II unstable, III unstable; if IV exists, it is stable

II stable, III unstable; IV cannot exist

II unstable, III stable; IV cannot exist

II stable, III stable; IV exists, but is unstable

† (e) Sketch the phase portrait for $A \geqslant 0$, $B \geqslant 0$ in each of the four cases of (d).

(f) Discuss the implications of the results for the population interaction of (a).

Appendix 6

Complex numbers

Complex numbers are of the form $a+ib$, where $i^2=-1$. Here a and b are ordinary (real) numbers, called, respectively, the **real** and the **imaginary parts** of the complex number.

Two complex numbers are regarded as equal if and only if their real and imaginary parts are equal. If

$$z_1=a_1+ib_1 \quad \text{and} \quad z_2=a_2+ib_2,$$

then

$$z_1=z_2 \quad \text{if and only if } a_1=a_2, \ b_1=b_2.$$

Arithmetic operations are defined in a natural way:

$$z_1+z_2=(a_1+a_2)+i(b_1+b_2), \tag{1}$$

$$z_1-z_2=(a_1-a_2)+i(b_1-b_2), \tag{2}$$

$$z_1 z_2=(a_1+ib_1)(b_2+ib_2)=a_1 a_2-b_1 b_2+i(b_1 a_2+a_1 b_2), \tag{3}$$

$$\frac{z_1}{z_2}=\frac{a_1+ib_1}{a_2+ib_2}=\frac{a_1+ib_1}{a_2+ib_2}\frac{a_2-ib_2}{a_2-ib_2}=\frac{a_1 a_2+b_1 b_2}{a_2^2+b_2^2}+i\,\frac{b_1 a_2-a_1 b_2}{a_2^2+b_2^2}. \tag{4}$$

Note that the sum, difference, product, and quotient of complex numbers are themselves complex numbers.

Elementary complex functions are also defined naturally. For example:

$$z^3=z\cdot z\cdot z, \quad z^{-1}=\frac{1}{z}.$$

Other complex functions are defined by using the same series that hold for real functions. For example [compare (A2.11), (A2.12), and (A2.10)]:

$$\sin z=z-\frac{z^3}{3!}+\dots, \tag{5}$$

$$\cos z=1-\frac{z^2}{2!}+\dots, \tag{6}$$

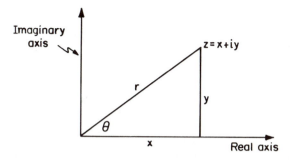

Figure A6.1. Representation in the complex plane of $z = x + iy = re^{i\theta}$.

$$e^z = 1 + z + \frac{z^2}{2!} + \dots . \tag{7}$$

It can be shown that the rules for the real functions carry over for their complex counterparts. For example:

$$e^{z_1 + z_2} = e^{z_1} e^{z_2}, \quad \sin(z_1 + z_2) = \sin z_1 \cos z_2 + \cos z_1 \sin z_2. \tag{8}$$

The derivative is defined as for real functions:

$$\frac{df(z)}{dz} = \lim_{h \to 0} \frac{f(z+h) - f(z)}{h}.$$

This must hold for any complex number h that tends to the complex zero $0 + i0$. The remarkable fact can be shown that definitions such as (5)–(7) are essentially the only possible definitions of $\sin z$, $\cos z$, and so forth, that are differentiable and that agree with the original definitions when z is real.

An important result is **de Moivre's Theorem** for e^{iy}, y real:

$$e^{iy} = \cos y + i \sin y. \tag{9}$$

This follows from (7), (6), and (5):

$$e^{iy} = 1 + iy + \frac{(iy)^2}{2!} + \frac{(iy)^3}{3!} + \frac{(iy)^4}{4!} + \frac{(iy)^5}{5!} + \cdots$$

$$= 1 - \frac{y^2}{2!} + \frac{y^4}{4!} - \cdots + i\left(y - \frac{y^3}{3!} + \frac{y^5}{5!} - \cdots \right).$$

The complex number $z = x + iy$ can be represented as a point in the x-y **complex plane** (Figure A6.1). It is convenient to introduce polar coordinates

$$x = r \cos \theta, \quad y = r \sin \theta, \tag{10}$$

so that

$$x^2 + y^2 = r^2, \quad \tan \theta = y/x. \tag{11}$$

We can write

$$z = x + iy = r \cos \theta + ir \sin \theta = r(\cos \theta + i \sin \theta).$$

Using de Moivre's theorem (9), we obtain

$$z = re^{i\theta}. \tag{12}$$

Because the laws of exponentiation hold, this **polar representation** of complex numbers provides an easy way to multiply them. Thus, if

$$z_1 = r_1 e^{i\theta_1}, \quad z_2 = r_2 e^{i\theta_2}, \quad \text{then } z_1 z_2 = r_1 r_2 e^{i(\theta_1 + \theta_2)}. \tag{13}$$

In particular, for any integer n,

$$z^n = (re^{i\theta})^n = r^n e^{in\theta}. \tag{14}$$

Exercises

†   1. If $z_1 = 7 - 3i$ and $z_2 = 1 + 2i$, write the following in the form $x + iy$ (do not substitute into formulas; calculate directly):

$$z_1 + z_2, \quad z_1 - 2z_2, \quad z_1 z_2, \quad z_1^2; \quad z_1/z_2.$$

†   2. Write the following in the form $x + iy$:

$$e^{i\pi}, \quad e^{i\pi/4}, \quad 2e^{i\pi/3}.$$

†   3. (a) Write the following in the form $re^{i\theta}$:

$$(\sqrt{2})(1 + i), \quad \frac{1}{2} + i\frac{\sqrt{3}}{2}.$$

      (b) Write $[(\sqrt{2})(1 + i)]^{10}$ and $[\frac{1}{2} + i(\sqrt{3})/2]^{75}$ in the form $x + iy$.

     4. The **modulus** and the **argument** of a complex number $z = x + iy$ are defined by

$$|z| = (x^2 + y^2)^{1/2}, \quad \arg z = \tan^{-1}(y/x) \quad [\text{i.e., } \tan z = y/x]. \tag{15}$$

      (a) Show that

$$|z_1 \cdot z_2| = |z_1||z_2|, \quad \arg(z_1 z_2) = \arg z_1 + \arg z_2. \tag{16}$$

      (b) Show that if $z = x + iy$, then $|e^z| = e^x$.

     5. The **complex conjugate** of $z = x + iy$, written $\bar{z}$ (or sometimes z^*), is defined by

$$\bar{z} = x - iy.$$

      Show that

$$\overline{re^{i\theta}} = re^{-i\theta}, \quad |z|^2 = z\bar{z}. \tag{17}$$

Appendix 7

Dimensionless variables

The use of dimensionless variables is a relatively easy way to rearrange a mathematical model so that the number of parameters that appear is reduced. The method provides advantages for the theoretician and experimentalist alike. We illustrate it here on a simple example. Several other examples are found in the main text.

Dimerization: a sample problem

Let us consider a "dimerization" reaction in which two molecules A reversibly combine to form a complex C. The reaction is symbolized by

$$A + A \underset{k_{-1}}{\overset{k_1}{\rightleftharpoons}} C.$$

According to the law of mass action (see Chapter 4), the reaction equations are

$$dA/dt = -2k_1 A^2 + 2k_{-1} C, \tag{1a}$$

$$dC/dt = k_1 A^2 - k_{-1} C. \tag{1b}$$

In (1a) we have taken account of the fact that each dimerization results in the loss of two A molecules, and each breakup of the dimer C results in the reappearance of two A molecules.

We shall consider an initial situation in which all the molecules are in the free form, at some concentration A_0:

$$\text{At } t = 0, \ A = A_0 \quad \text{and} \quad C = 0. \tag{2}$$

The differential equations (1) and the initial conditions (2) constitute the mathematical problem that we wish to discuss.

Before proceeding with the matter at hand, we note that (1a) and (1b) can be combined to yield

$$\frac{dA}{dt} + 2\frac{dC}{dt} = 0, \tag{3}$$

275

or

$$\frac{d}{dt}(A+2C)=0, \quad A+2C=\text{constant}=A_0. \tag{4}$$

This result is indeed correct, and it thus provides a check on our derivation. Since no molecules are destroyed, they are always found in a free form (A molecules per unit volume) or in the form of the complex ($2C$ molecules per unit volume – because each C complex is formed of two A molecules).

Three parameters appear in (1) and (2): k_{-1}, k_1, and A_0. Recording the accumulation of complex C with time would consequently appear to require a whole "encyclopedia" of graphs. On one page we could draw graphs of C as a function of t for several different values of k_{-1}, keeping k_1 and A_0 fixed. On another page we could present the same type of graphs, but for a different value of k_1 – and the same value of A_0. Thus, in a book of graphs, we could display the data for the dependence of C on t, k_{-1}, and k_1, for a fixed value of A_0. A set of books, one for each A_0, would allow complete presentation of the results for all possible situations.

Introducing dimensionless variables

We now explain how the introduction of dimensionless variables shows that the full set of results can be presented on a single page. We present a general step-by-step procedure for the introduction of dimensionless variables and immediately illustrate it on the specific example at hand.

1. *Determine the dimensions of each parameter and variable in the problem.* The parameter A_0 in (2) represents a concentration – number of molecules per unit volume (e.g., the number per cm^3). In most problems, the units of mass (M), length (L), and time (T) are sufficient to express the dimensions. Volume is measured as the cube of some length, so that a concentration has dimension $(\text{length})^{-3}$. It is conventional to use the symbol "[ ]" to mean "the dimension of," so that we write

$$[A_0]=L^{-3}; \quad \text{also,} \quad [A]=L^{-3}, \quad [C]=L^{-3}. \tag{5}$$

The rate constant k_{-1} gives the fraction of complex molecules that break up per unit time, so that

$$[k_{-1}]=T^{-1}. \tag{6}$$

Of course,

$$[t] = T. \tag{7}$$

To determine the dimensions of k_1, we can use the fact that *each term in a properly posed equation must have the same dimensions.* Otherwise the ridiculous situation would result wherein different terms would change in different ways if new units were introduced. For example, if one term had dimensions L^2 and the second T, if centimeters were used instead of meters, then the first term would be multipled by 10^4, while the second term would not be altered.

In the differential equation (1a), each term must have dimensions of concentration/time (i.e., of $L^{-3}T^{-1}$), because

$$[k_{-1}C] = [k_{-1}][C] = T^{-1}L^{-3}. \tag{8}$$

But

$$[k_1 A^2] = [k_1][A]^2 = [k_1]L^{-6},$$

so that, from (8),

$$[k_1]L^{-6} = T^{-1}L^{-3}, \quad [k_1] = L^3 T^{-1}. \tag{9}$$

This completes Step 1.

2. *Introduce new dimensionless dependent and independent variables by dividing by any suitable combination of parameters.* A quantity is called **dimensionless** if its value is independent of units. Thus, to define a dimensionless version τ of the time t, we can take

$$\tau = \frac{t}{1/k_{-1}} = k_{-1}t. \tag{10}$$

The variable τ is the ratio of the actual time to the time $1/k_{-1}$. For example, if t is 3 sec and k_{-1} is 10^3 sec^{-1}, then $\tau = (10^3 \text{ sec}^{-1})(3 \text{ sec}) = 3 \cdot 10^3$. The ratio τ is unaltered if minutes are used instead of seconds:

$$\tau = 0.05 \text{ min}, \quad k_{-1} = 6 \cdot 10^4 \text{ min}^{-1},$$

$$\tau = (0.05 \text{ min})(6 \cdot 10^4 \text{ min}^{-1}) = 3 \cdot 10^3. \tag{11}$$

To show formally that τ is dimensionless, we use (10), (6), and (7) to write

$$[\tau] = [k_{-1}t] = [k_{-1}][t] = T^{-1}T = T^0.$$

From (9) and (5), $(A_0 k_1)^{-1}$ also has the dimension of time, and this combination of parameters could have been used instead of $1/k_{-1}$ in the definition of a dimensionless time. The same type of result emerges no matter how the dimensionless variables are chosen (Exercise 1).

From (5), an obvious choice for dimensionless concentrations a and c is

$$a = \frac{A}{A_0}, \quad c = \frac{C}{A_0}. \tag{12}$$

3. *Rewrite the equations and boundary conditions in the new variables.* In our example, the relationships between the original dimensional variables C, A, and t and the new dimensionless variables c, a, and τ are as follows:

$$C = A_0 c, \quad A = A_0 a, \quad t = \tau / k_{-1}. \tag{13}$$

Substituting from (13), we easily find that the initial conditions (2) become the following:

$$\text{At } \tau = 0, \, a = 1 \quad \text{and} \quad c = 0. \tag{14}$$

The terms on the right side of the differential equations (1) are readily written in terms of the new variables, but the derivatives at first present a problem. The rigorous way to proceed is via the chain rule (A1.10). For example:

$$\frac{dA}{dt} = \frac{d(A_0 a)}{dt} = A_0 \frac{da}{dt} = A_0 \frac{da}{d\tau} \frac{d\tau}{dt} = A_0 \frac{da}{d\tau} k_{-1} = k_{-1} A_0 \frac{da}{d\tau}. \tag{15}$$

Thus, (1a) becomes

$$k_{-1} A_0 \frac{da}{d\tau} = -2 k_1 A_0^2 a^2 + 2 k_{-1} A_0 c. \tag{16}$$

It is best to *simplify the dimensionless equations by making one of the coefficients unity.* In the present case, we do this by dividing both sides of (16) by $k_{-1} A_0$. This yields

$$\frac{da}{d\tau} = -2\phi a^2 + 2c, \tag{17}$$

where we have made the definition

$$\phi \equiv k_1 A_0 / k_{-1}. \tag{18}$$

Note that the result of (15) could have been obtained if we had naively performed the substitutions (13) and transferred the constants outside the derivative:

$$\frac{dA}{dt} = \frac{d(A_0 a)}{d(\tau / k_{-1})} = k_{-1} A_0 \frac{da}{d\tau}. \tag{19}$$

This "naive" method is recommended in practice.

As the reader should verify, substitution of (13) into (1b) yields

$$\frac{dc}{d\tau} = \phi a^2 - c. \tag{20}$$

Thus, in our new variables, the mathematical problem (1a), (1b), and (2) is transformed into (17), (20), and (14). Only a single parameter ϕ now appears. This parameter is dimensionless, for (18), (9), (5), and (6) imply

$$[\phi] = \frac{[k_1][A_0]}{[k_{-1}]} = \frac{L^3 T^{-1} \cdot L^{-3}}{T^{-1}} = L^0 T^0. \tag{21}$$

It would be astonishing if ϕ were not dimensionless, for the other terms in equations (17) and (20) are dimensionless.

Advantages of dimensionless variables

We have completed the task of introducing dimensionless variables. The conclusions observed here are general. *When the new equations are written in their simplest form, they contain only dimensionless variables and dimensionless parameters. The number of dimensionless parameters is generally smaller than the original number of parameters.*

The reduction in the number of parameters makes theoretical manipulations easier, for the equations are less cluttered. Often more important is the fact that the use of dimensionless variables permits results to be obtained and displayed far more economically. In the present case, we pointed out that a shelf of books appeared to be necessary to present the complex concentration C as a function of the time t and of the parameters k_{-1}, k_1, and A_0. We now see from the equations that the dimensionless complex concentration c depends only on the single dimensionless parameter ϕ. In other words, the variable combination C/A_0 is not a general function of the parameters k_{-1}, k_1, and A_0 and the time t but rather a function only of the combinations $k_1 A_0/k_{-1}$ and $k_{-1}t$. Thus, all the data concerning dimerization can, in principle, be presented on a single page, a graph of $c \equiv C/A_0$ as a function of $\tau \equiv k_{-1}t$ for a number of different values of $\phi \equiv k_1 A_0/k_{-1}$ (Figure A7.1).

We stress that our conclusions about the reduction in parameter number were obtained without solving any equations. Such conclusions can thus be of great importance for purely experimental (or numerical) work, in minimizing the amount of experimentation that is necessary to describe the possible variation of the results for different parameter values.

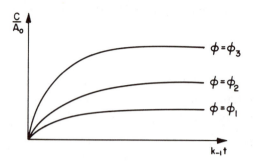

Figure A7.1. Schematic graphs showing several solutions C to (1) and (2), for various values ϕ_i of the dimensionless parameter $\phi \equiv k_1 A_0 / k_{-1}$. [Equivalently, the graphs depict solutions to the dimensionless problem given by (14), (17), and (20).]

Nondimensionalizing a functional relationship

We close our discussion here by illustrating how our conclusions could have been reached without the use of any equations whatever. This approach to the subject will not be used elsewhere in the text, but it deepens understanding of the material. [For further discussion of dimensionless variables, see, e.g., Lin and Segel (1974) and the references cited therein.]

Suppose that our grasp of the dimerization process is sufficient to permit the conclusion that the amount of complex C can depend only on the time t and the parameters k_{-1}, k_1, and A_0. Consider the dimensionless ratio C/A_0. This quantity can depend only on dimensionless combinations of t, k_{-1}, k_1, and A_0. For example, an equation such as

$$\frac{C}{A_0} = k_1 + A_0^2$$

must be incorrect, for the left side is dimensionless, but the right side is altered if units are changed. The most general possible dimensionless combination of t, k_1, k_{-1}, and A_0 must take the form

$$P \equiv t^a k_1^b k_{-1}^c A_0^d, \tag{22}$$

where a, b, c, and d are constants. Inserting the known dimensions, we find

$$[P] = T^a (L^3 T^{-1})^b (T^{-1})^c (L^{-3})^d = T^{a-b-c} L^{3b-3d}. \tag{23}$$

If P is to be dimensionless, it must be that

$$a - b - c = 0, \quad 3b - 3d = 0. \tag{24}$$

Thus, c and d can be arbitrary (to take one of several equivalent possibilities), but

$$b=d, \quad a=b+c; \quad \text{i.e.,} \quad a=d+c. \tag{25}$$

Consequently, from (22),

$$P=t^{d+c}k_1^d k_{-1}^c A_0^d = (tk_1 A_0)^d (k_{-1}t)^c. \tag{26}$$

We conclude that the dimensionless combinations of t, k_1, k_{-1}, and A_0 on which C/A_0 depends must be functions of $tk_1 A_0$ and $k_{-1}t$. Indeed, we chose for our parameters

$$k_{-1}t \equiv \tau, \quad \frac{tk_1 A_0}{k_{-1}t} = \frac{k_1 A_0}{k_{-1}} \equiv \phi. \tag{27}$$

Exercises

1. Find the dimensionless version of (1) and (2) when, instead of (10), the dimensionless time

   $$\tau_1 = k_1 A_0 t$$

   is employed. Show that the dimensionless equations still depend only on a single dimensionless parameter. (This illustrates the fact that, in general, the same number of dimensionless parameters appear, independent of how the dimensionless variables are chosen.) In the present case, whether the forward rate $k_1 A_0$ or the backward rate k_{-1} is selected for a dimensionless time, the proportionate values of the concentrations depend only on the ratio of these rates.

2. (a) Find steady-state solutions $\bar{A}$ and $\bar{C}$ of (1) and (4).
   (b) Show that $\bar{A}/A_0$ and $\bar{C}/A_0$ [where $\bar{A}$ and $\bar{C}$ are given in (a)] depend on k_1, k_{-1}, and A_0 only in the combination $k_1 A_0/k_{-1}$.
   (c) As a check, find the steady-state solutions of (17), (20), and the dimensionless counterpart of (4). Compare with the results of (b).
   (d) The formal calculations yield two possible steady-state solutions. Show that (4), or its dimensionless counterpart, implies that only the larger of these solutions is chemically meaningful.

3. (a) If you are familiar with the method of separation of variables (and integration by partial fractions) solve (1) and (2). Begin by employing (4) to eliminate one of the variables.
   (b) Rearrange the solution of (a) to show that $A(t)/A_0$ and $C(t)/A_0$ depend only on τ and ϕ.
   (c) As a check, directly find the solution to (17), (20), and (14), and compare the results with (b).

Appendix 8

Integration

To motivate the definition of the integral, consider the problem of finding the area under the curve $y = f(x)$, f positive, between $x = a$ and $x = b$. This area lies between the sums

$$\sum_{i=1}^{N} (f_i^{(m)} \Delta x) \quad \text{and} \quad \sum_{i=1}^{N} (f_i^{(M)} \Delta x), \tag{1}$$

where

$$\Delta x \equiv (b - a)/N.$$

Here, $f_i^{(m)}$ ($f_i^{(M)}$) is the smallest (largest) value of $f(x)$ for x between $a + (i-1)(\Delta x)$ and $a + i(\Delta x)$. Moreover, if Δx is small, the area can be approximated by the sum

$$\sum_{i=1}^{N} (f(x_i^*) \Delta x),$$

where x_i^* is any point between $a + (i-1)\Delta x$ and $a + i(\Delta x)$ (Figure A8.1). If f is continuous for $a \leqslant x \leqslant b$, then it can be shown that all three sums approach the same limit as $N \to \infty$. This limit is called the **integral of f between a and b** and is written $\int_a^b f(x)\, dx$.

The integral has many other interpretations, in addition to that as an area. For example, consider a cylindrical tube of constant cross-sectional area A. Suppose that the tube contains a chemical whose concentration C does not vary radially but does change as a function of axial distance x, $a \leqslant x \leqslant b$. A thin slice of the tube between x_i and $x_i + \Delta x$ has volume $A\Delta x$, and thus contains approximately $C(x_i^*)A\Delta x$ molecules. The total number of molecules in the tube is well approximated by a sum of the form

$$A \sum_i C(x_i^*) \Delta x.$$

In the limit as the slices become infinitely thin ($\Delta x \to 0$), we obtain the "exact" formula

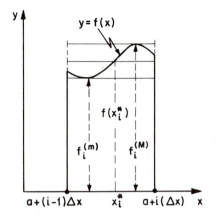

Figure A8.1. Illustration of quantities needed in the definition of the integral.

$$A \int_a^b C(x)\,dx \qquad\qquad (2)$$

for the number of molecules in the tube.

In calculus courses, we learn basic properties of integration such as

◆◆
$$\int_a^b Cf(x)\,dx = C\int_a^b f(x)\,dx, \quad C \text{ a constant,} \qquad (3)$$

◆◆
$$\int_a^b f(x)\,dx = -\int_b^a f(x)\,dx, \qquad\qquad (4)$$

◆◆
$$\int_a^b f(x)\,dx + \int_b^c f(x)\,dx = \int_a^c f(x)\,dx, \qquad (5)$$

and

◆◆
$$\int_a^b f(x)\,dx = F(b) - F(a), \qquad\qquad (6a)$$

if

$$\frac{dF(x)}{dx} = f(x). \qquad\qquad (6b)$$

Formula (6a) is often written in the form

$$\int_a^b f(x)\,dx = F(x)\Big|_a^b, \qquad\qquad (6c)$$

because, by definition,

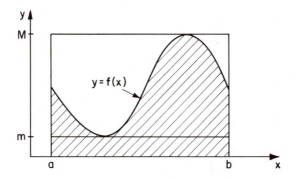

Figure A8.2. The area of the shaded region is greater than $m(b-a)$ and less than $M(b-a)$.

$$F(x)\bigg|_a^b \equiv F(b)-F(a). \tag{6d}$$

Example: Find $\int_1^3 7x^3\,dx$.
Solution: Because

$$\frac{d}{dx}\frac{x^4}{4}=x^3,$$

we have

$$\int_1^3 7x^3\,dx=7\int_1^3 x^3\,dx=7\cdot\frac{x^4}{4}\bigg|_1^3=7\left(\frac{3^4}{4}-\frac{1^4}{4}\right)=140. \tag{7}$$

The **fundamental theorem of calculus** is

◆◆ $$\frac{d}{dx}\int_a^x f(t)\,dt=f(x), \tag{8}$$

where a is any constant and f is continuous.

We now derive a useful theoretical result called the **integral mean-value theorem**. To this end we first note that the area interpretation of the integral (Figure A8.2) justifies the inequalities

$$m(b-a)\leqslant\int_a^b f(x)\,dx\leqslant M(b-a). \tag{9}$$

Here, M and m are the maximum and minimum values of $f(x)$ for $a\leqslant x\leqslant b$. Let us define I by

$$I\equiv\frac{1}{b-a}\int_a^b f(x)\,dx.$$

Then (9) implies that I is between m and M. But if f is continuous, there

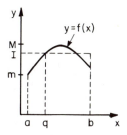

Figure A8.3. If I is between the minimum m and the maximum M of f, and f is continuous, then there is a q between a and b such that $f(q) = I$.

is at least one number q between a and b such that $I = f(q)$ (Figure A8.3). Consequently,

$$♦♦ \quad \int_a^b f(x)\, dx = f(q) \cdot (b - a), \quad \text{for some } q \text{ such that } a \leqslant q \leqslant b. \tag{10}$$

In order to afford a concrete illustration of the theorem, let $f(x)$ denote the spatially varying concentration of bacteria (number per unit length) that are confined between walls at $x = a$ and $x = b$. Then (10) has the interpretation that the total number of bacteria equals a typical concentration, at some point $x = q$, times the distance between the walls.

Two other important formulas (that we employ only in the supplement to Chapter 7) are given in the following paragraphs.

Often we wish to change the variable of integration, say from the given variable x to a new variable z, where x is a function of z described by the equation $x = g(z)$. This is accomplished by the **substitution formula**

$$♦♦ \quad \int_a^b f(x)\, dx = \int_A^B f[g(z)] \frac{dx}{dz}\, dz; \quad a = g(A),\ b = g(B). \tag{11}$$

If we introduce a function F of the special form $F(x) = g(x) h(x)$, we obtain, from (6b),

$$f(x) = g(x) \frac{dh(x)}{dx} + h(x) \frac{dg(x)}{dx}.$$

Substitution into (6a) and a minor rearrangement give the remarkably useful **integration-by-parts formula**

$$\int_a^b g(x) \frac{dh(x)}{dx}\, dx = g(b) h(b) - g(a) h(a) - \int_a^b h(x) \frac{dg(x)}{dx}\, dx. \tag{12}$$

With the notation of (6d), (12) can be written

$$\int_a^b g(x)\frac{dh(x)}{dx}\,dx = g(x)h(x)\Big|_a^b - \int_a^b h(x)\frac{dg(x)}{dx}\,dx. \tag{13}$$

Exercises

†   1. (a) $\displaystyle\int_2^3 (x+7)\,dx.$ (b) $\displaystyle\int_0^{\pi/2} \sin x\,dx.$ (c) $\displaystyle\int_0^{\pi/2} \cos 2x\,dx.$

        (d) $\displaystyle\int_2^3 \cos(x+3)\,dx.$

†   2. Use integration by parts to evaluate the following:

        (a) $\displaystyle\int_0^1 xe^x\,dx.$ (b) $\displaystyle\int_0^{\pi} x^2 \sin x\,dx.$

†   3. $\int_{\pi/2}^{\pi} \sin^3 x \cos x\,dx.$

†   4. Find a value of q such that (10) holds for the integral of Exercise 1(a).

        5. (a) If $F(x) = \int_1^x (t+t^2)\,dt$, find an explicit expression for $F(x)$ by carrying out the integration.

        (b) Differentiate the explicit expression for $F(x)$ in part (a) to obtain $dF(x)/dx = x + x^2$, thereby verifying (8) in a particular case.

†   6. $\displaystyle\frac{d}{dx}\int_a^{x^2} \sqrt{(t+1)}\,\sin t\,dt.$

Hints and answers for selected exercises

1: Optimal strategies for the metabolism of storage materials

    1.10 (a) There is a minimum at $x=0$. There is an additional minimum at $x=b$ if $b>1$.

    1.12 (c) *Hint:* To demonstrate that $\ln x + 2 - 2x < 0$ for $x \geqslant 1$, sketch on one set of axes the graphs of $y = \ln x$ and $y = 2(x-1)$.

2: Recursion relations in ecological and cellular population dynamics

    2.1 (b) *Hint:* The graph of $y = x^2 - x + r^{-1}$ intersects the x-axis only if there are real roots to the quadratic equation $x^2 - x + r^{-1} = 0$.

    2.2 (b) The four-year cycle is approximately 0.501, 0.875, 0.383, 0.827.

    2.3 Steady states are $x=0$ (monotonically stable), $x=1$ (monotonically unstable), and $x=3$ (oscillatorily stable). The population should tend to either $x=0$ or $x=3$, depending on whether the initial population x_0 is less than or greater than unity.

    2.4 (b) *Hint:* As a first step in linearizing the equation for $y_i = x_i - x$, use Taylor series to linearize $\exp[s(1 - y_{i-1} - x)]$. *Answer:* The steady state $x=0$ is always unstable, whereas $x=1$ is stable for $0 < s < 2$.

    2.5 (f) $y_p = (C_1 \cos p\theta + C_2 \sin p\theta)(r-1)^{p/2}$, where C_1 and C_2 are arbitrary constants; $\tan \theta = \sqrt{(4r-5)}$.

    2.7 (c) The magnitude of the slope must be less than unity.

    2.8 (c) $\bar{x}$ exists if $f'(0) > 1$.

    (d) *Hint:* By the chain rule,

$$\frac{df_2(x)}{dx} = \frac{df(y)}{dy}\bigg|_{y=f(x)} \frac{df(x)}{dx}.$$

    (e) *Hint:* Demonstrate and use the relation

$$\frac{d^2 f_2(x)}{dx^2} = \frac{df(y)}{dy}\bigg|_{y=f(x)} \frac{d^2 f(x)}{dx^2} + \left[\frac{df(x)}{dx}\right]^2 \frac{d^2 f(y)}{dy^2}\bigg|_{y=f(x)}$$

Here, $f'(\bar{x}) < -1$.

4: Enzyme kinetics

    4.10 (c) The reason that positive cooperativity is not necessarily associated with an increasing first derivative of $Y(z)$ (concavity) must be that effects of saturation are relatively important even at very low concentrations. One would conjecture that a region of concavity is sufficient but not necessary to conclude the presence of positive cooperativity.

5: The chemostat

5.5 *Hint:* Use Taylor series, remembering that $k(0)=r(0)=0$.

5.11 (a) *Partial answer:* Bacteria grow exponentially at a rate proportional to their consumption rate of some limiting chemical of concentration $C(t)$, as in the chemostat. A constant death rate d is assumed, as is a constant input rate I of limiting chemical.

5.12 *Partial answer:* Equation (2) might take the form

$$N(t+\Delta t)=N(t)+k[C(t)]N(t)\Delta t-qN(t)\Delta t.$$

Such a discrete model involves no calculus and is suitable for direct calculation on a computer.

7: Diffusion

7.1 According to equation (17), the ratio of motilities is given by

$$\frac{D_1}{D_2}=\left(\frac{N_1}{N_2}\right)^2\frac{T_2}{T_1}=\left(\frac{3}{11}\right)^2\frac{3}{2}\approx0.1.$$

7.3 (a) *Hint:* By the chain rule,

$$\frac{\partial u(x,t)}{\partial t}=\frac{df(z)}{dz}\bigg|_{z=x-ct}\frac{\partial z}{\partial t}=-c\left[\frac{df(z)}{dz}\right]_{z=x-ct}.$$

(e) *Hint:* Impose the initial conditions. Differentiate one of the resulting equations, and obtain two equations for the two unknowns $f'(x)$ and $g'(x)$ (where $'\equiv d/dx$). *Answer:* $u=\frac{1}{2}F(x-ct)+\frac{1}{2}F(x+ct)$.

(f) In this case, and in general, the solution is composed of two identically shaped waves, each having half the amplitude of the original $F(x)$, and each moving at speed c, one to the right and one to the left.

8: Developmental pattern formation and stability theory

8.2 In (50a), the terms multiplied by D_a' are replaced by $a_{i+1,j}+a_{i-1,j}+a_{i,j+1}+a_{i,j-1}-4a_{i,j}$.

8.3 (a) $R=6\frac{2}{3}$, $R_c=2.1$.

(b) If we take $\rho_a=\rho_h=0.5$, say, then, $R=66\frac{2}{3}$. This value is so much larger than $R_c=2.04$ that strong instability seems assured, even though the varying secretion rates are replaced by an estimate of their average values.

8.5 $p_n=C_n\lambda^{-1}\int_0^\lambda p(s)\cos[n\pi(s/\lambda)]\,ds$, $q_n=C_n\lambda^{-1}\int_0^\lambda q(s)\cos[n\pi(s/\lambda)]\,ds$; $C_n=2$, $n\geq1$; $C_0=1$.

8.9 (c) *Hint:* Use the chain rule and discuss the case $l=0$ separately.

9: A mechanical basis for morphogenesis

9.2 (a) *Hint:* To ascertain the number of zeros of dL/dC, work with graphs of $S(C)$, both when dS/dC at the origin is positive and when it is zero.

9.3 A sketch is given below.

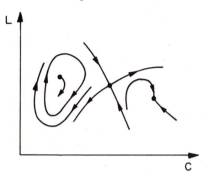

Appendix 1: Mathematical prerequisites

A1.1 (a) $3x^2$. (b) $-2x^{-3}$. (c) $-5x^{-6}=-5/x^6$. (d) $2x+3.5x^{2.5}$.

(e) $-21x^{-4}$. (f) $12x^2$. (g) $-6x^{-4}$. (h) $-2x^{-3}-3x^2$.

A1.2 (a) $x\cos x+\sin x$. (b) $(\sin x-x\cos x)/\sin^2 x$ or $(\csc x)(1-x\cot x)$.

(c) $-2\sin 2x$. (d) $-2x\sin(x^2)$.

A1.3 (a) $\dfrac{d}{dx}\sec x=\dfrac{d}{dx}(\cos x)^{-1}=-1(\cos x)^{-2}(-\sin x)=\tan x\sec x$.

(b) $-\csc x\cot x$. (c) $\sec^2 x$. (d) $-\csc^2 x$.

A1.4 (a) $2x\exp(x^2)$. (b) $\cot x$. (c) $(x^{-1}-\ln x)/e^x$. (d) $15x^2\exp(5x^3)$.

A1.5 (a) $-x/y$. (b) $-1/x$. (c) $y'=[2xy(x+y)-1]/[1-x^2(x+y)]$.

(d) $-3\cos 3x/[2\sin 2y]$.

A1.6 (a) $30x^2y^2$. (b) $2zxye^{xyz^2}$. (c) $ze^{zt}\cos(e^{zt})$.

Appendix 2: Infinite series and Taylor approximations

A2.2 (a) $a_0=1$, $a_1=0$, $a_2=-\frac{1}{2}$.

Appendix 3: Difference equations

A3.1 (a) -125. (b) $x_n=4(\frac{1}{3})^{n-1}=4(3)^{1-n}$. (c) $z_k=(-\pi)^k$.

A3.2 (b) *Hint:* Show that $F=1+F^{-1}$.

A3.4 (a) $z_n=C_1+C_2\cdot 2^n$. (b) $z_n=-2+2^n$.

A3.5 (a) $k=2\pm 3i$. (c) $C_1=1$, $C_2=(1-\cos\theta)/\sin\theta$.

Appendix 4: Linear differential equations with constant coefficients

A4.1 (a) $x=Ce^{7t}$ and $y=Ce^{2t}$, (b) $z=2e^{-2t}$.

(c) $x=C_1e^{-3t}+C_2e^{-4t}$ and $y=C_1e^{t/2}+C_2e^{-5t}$.

(d) $y=\frac{1}{2}(-e^{-3t}+e^t)$.   (e) $z=-2e^{x/2}+3e^{x/3}$.

A4.2 $-2<p<2$; $y=\exp(-p/2)[C_1\cos\sqrt{(1-\frac{1}{4}p^2)}+C_2\sin\sqrt{(1-\frac{1}{4}p^2)}]$.

A4.4 (c) In (a), $C_1=3$ and $C_2=2$.

Appendix 5: Phase-plane analysis

A5.5 *Hint:* The result can be obtained from Figure A5.9.

A5.7 In Figure A5.13(a), A and C are stable, and B is unstable. Solutions tend to A or C, depending on initial conditions. In Figure A5.13(b), the steady state is stable. (A closed curve corresponds to a possible periodic solution, but it is unstable.) Trajectories tend to the steady state if they start within the closed curve. It is not clear from the figure whether trajectories that begin outside the closed curve eventually meet the axes (zero concentration) or spiral off to infinity.

A5.10 (a) *Partial answer:* The terms $x(a-bx)$ give the birth of the prey, with a birthrate that equals a at low prey concentrations, but decreases linearly as prey concentration increases.

A5.13 (h) $\tau=tq$, $v=V/W$, $s=aS$.

A5.14 (a) *Partial answer:* The equations could describe a situation of competition between A and B, because the growth rate of A is reduced when B becomes large, and vice versa. The form of the term $-a_2AB^2$ is unusual, as it seems to indicate that the cooperation of two B individuals is involved in the reduction of the A population.

(e) Some of the phase portraits are depicted in the supplement to Chapter 8.

Appendix 6: Complex numbers

A6.1 $8-i$, $5-7i$, $13+11i$, $40-42i$, $\dfrac{1}{5}-\dfrac{17}{5}i$.

A6.2 $-1+i\cdot 0=-1$, $\dfrac{1}{\sqrt{2}}+\dfrac{i}{\sqrt{2}}$, $1+i\sqrt{3}$.

A6.3 (a) $2e^{i\pi/4}$, $e^{i\pi/3}$.

(b) $2^{10}e^{i10\pi/4}=2^{10}\left(\cos\dfrac{5\pi}{2}+i\sin\dfrac{5\pi}{2}\right)=2^{10}i$, -1.

Appendix 8: Integration

A8.1 (a) 19/2.   (b) 1.   (c) $\frac{1}{2}$.   (d) $\sin 6-\sin 5$.

A8.2 (a) 1.   (b) $[(2-x^2)\cos x+2x\sin x]_0^{\pi/2}=\pi-2$.

A8.3 $\int_1^0(u^3/3)\,du=-\frac{1}{4}$.

A8.4 $q=\frac{5}{2}$.

A8.6 *Hint:* Set $u=x^2$, and use the chain rule to obtain $2x(x^2+1)^{1/2}\sin(x^2)$.

References

Adler, J. (1969). Chemoreceptors in bacteria. *Science,* 166:1588–97.

Bard, J. B. L. (1978). A quantitative model of liver regeneration in the rat. *J. Theoret. Biol.,* 73:509–30.

– (1979). A quantitative theory of liver regeneration in the rat. II. Matching an improved mitotic inhibitor model to the data. *J. Theoret. Biol.,* 79:121–36.

Birkhoff, G., and Rota, G.-C. (1962). *Ordinary Differential Equations.* Boston: Ginn & Co.

Bonner, J. T. (1967). *The Cellular Slime Molds,* 2nd ed. Princeton University Press.

Bossert, W. H., and Wilson, E. O. (1963). The analysis of olfactory communication among animals. *J. Theoret. Biol.,* 5:443–69.

Boyce, W. E., and DiPrima, R. C. (1977). *Elementary Differential Equations and Boundary Value Problems,* 3rd ed. New York: Wiley.

Briggs, G. E., and Haldane, J. B. S. (1925). A note on the kinetics of enzyme action. *Biochem. J.,* 19:338–9.

Child, C. M. (1941). *Patterns and Problems of Development.* University of Chicago Press.

Cohen, D., and Parnas, H. (1976). An optimal policy for the metabolism of storage materials in unicellular algae. *J. Theoret. Biol.,* 56:1–18.

Cohen, M. S. (1977). The cyclic AMP control system in the development of *Dictyostelium discoideum. J. Theoret. Biol.,* 69:57–85.

Condon, E., and Shortley, F. (1957). *The Theory of Atomic Spectra.* Cambridge University Press.

Cooke, K., and Grossman, Z. (1982). Discrete delay, distributed delay and stability switches. *J. Math. Anal. Appl.,* 86:592–627.

Coukell, M. B., and Chan, F. K. (1980). The precocious appearance and activation of an adenylate cyclase in a rapidly developing mutant of *Dictyostelium discoideum. F.E.B.S. Letters,* 110:39–42.

Coullet, P. H., and Spiegel, E. A. (1983). Amplitude equations for systems with competing instabilities. *S.I.A.M. J. Appl. Math.,* 43:776–821.

Crank, J. (1975). *The Mathematics of Diffusion,* 2nd ed. Oxford: Clarendon Press.

Cushing, J., and Saleem, M. (1982). A predator prey model with age structure. *J. Math. Biol.,* 14:231–50.

Darmon, M., Barra, J., and Brachet, P. (1978). The role of phosphodiesterase in aggregation of *Dictyostelium discoideum. J. Cell Sci.,* 31:233–43.

DeSapio, R. (1978). *Calculus for the Life Sciences.* San Francisco: Freeman.

Devreotes, P. N., and Steck, T. L. (1979). Cyclic 3′,5′AMP relay in *Dictyostelium discoideum.* Requirements for initiation and termination of the response. *J. Cell Biol.,* 80:300–9.

Dinauer, M. C., MacKay, S. A., and Devreotes, P. N. (1980). Cyclic 3′,5′-AMP relay in *Dictyostelium discoideum*. III. The relationship of cAMP synthesis and secretion during the cAMP signalling response. *J. Cell Biol.*, 86:537–44.

Dinauer, M. C., Steck, T. L., and Devreotes, P. N. (1980). Cyclic 3′,3′-AMP relay in *Dictyostelium discoideum*. V. Adaptation of the cAMP signalling response during cAMP stimulation. *J. Cell Biol.*, 86:554–61.

Durston, A. (1974). Pacemaker mutants of *Dictyostelium discoideum*. *Dev. Biol.*, 38:308–19.

Eigen, M., and Shuster, P. (1977, 1978). The Hypercycle. Parts A, B, and C. *Naturwissenschaften*, 64:541–65; 65:7–41, 341–69.

Eisen, M., and Schiller, J. (1977). Stability analysis of normal and neoplastic growth. *Bull. Math. Biol.*, 66:799–809.

Feigenbaum, M. (1980). Universal behavior in nonlinear systems. *Los Alamos Science*, 1:4–27.

Frisch, H. L., and Gotham, I. J., III (1977). On periodic algal cyclostat populations. *J. Theoret. Biol.*, 66:665–78.

Gerisch, G. (1968). Cell aggregation and differentiation in *Dictyostelium*. *Curr. Top. Dev. Biol.*, 3:157–97.

Gerisch, G., and Hess, B. (1974). Cyclic-AMP controlled oscillations in suspended *Dictyostelium* cells: Their relation to morphogenetic cell interactions. *Proc. Natl. Acad. Sci. U.S.A.*, 71:2118–22.

Gerisch, G., Maeda, Y., Malchow, D., Roos, W., Wick, U., and Wurster, B. (1977). Cyclic AMP signals and the control of cell aggregation in *Dictyostelium discoideum*. In: *Development and Differentiation in the Cellular Slime Molds*, edited by P. Cappuccinelli and J. M. Ashworth, pp. 105–24. Amsterdam: Elsevier/North Holland.

Gerisch, G., and Wick, U. (1975). Intracellular oscillations and release of cyclic AMP from *Dictyostelium* cells. *Biochem. Biophys. Res. Commun.*, 65:364–70.

Gierer, A., and Meinhardt, H. (1972). A theory of biological pattern formation. *Kybernetik*, 12:30–9.

Goldberg, S. (1961). *Introduction to Difference Equations*. New York: Wiley.

Goldbeter, A. (1980). Models for oscillations and excitability in biochemical systems. In: *Mathematical Models in Molecular and Cellular Biology*, edited by L. A. Segel, pp. 248–91. Cambridge University Press.

Goldbeter, A., Erneux, T., and Segel, L. A. (1978). Excitability in the adenylate cyclase reaction in *Dictyostelium discoideum*. *F.E.B.S. Letters*, 89:237–41.

Goldbeter, A., and Martiel, J. L. (1982). A critical discussion of plausible models for relay and oscillation of cyclic AMP in *Dictyostelium* cells. In: *Lecture Notes in Biomathematics, Vol. 49: Rhythms in Biology and Other Fields of Application*, edited by M. Cosnard, J. Demongeot, and A. Le Breton, pp. 173–88. Berlin: Springer-Verlag.

Goldbeter, A., and Segel, L. A. (1977). Unified mechanism for relay and oscillation of cyclic AMP in *Distyostelium discoideum*. *Proc. Natl. Acad. Sci. U.S.A.*, 74:1543–7.

– (1980). Control of developmental transitions in the cyclic AMP signalling system of *Dictyostelium discoideum*. *Differentiation*, 17:127–35.

Granero, M. I., Porati, A., and Zanacca, D. (1977). A bifurcation analysis of pattern formation in a diffusion governed morphogenetic field. *J. Math. Biol.*, 4:21–7.

Gressel, J., and Segel, L. A. (1978). The paucity of plants evolving genetic resistance to herbicides: Possible reasons and implications. *J. Theoret. Biol.,* 75:349–71.

Guevara, M. R., and Glass, L. (1982). Phase locking, period doubling bifurcations and chaos in a mathematical model of a periodically driven oscillator: A theory for the entrainment of biological oscillators and the generation of cardiac dysrhythmias. *J. Math. Biol.,* 14:1–24.

Guevara, M. R., Glass, L., and Shrier, A. (1981). Phase locking, period-doubling bifurcations and irregular dynamics in periodically stimulated cardiac cells. *Science,* 214:1350–2.

Hagan, P. S., and Cohen, M. S. (1981). Diffusion-induced morphogenesis in the development of Dictyostelium. *J. Theoret. Biol.,* 93:881–908.

Hainsworth, F. R., and Wolf, L. L. (1979). Seeding: An ecological approach. *Adv. Study Behav.,* 9:53–96.

Harris, A. K., Stopak, D., and Warner, P. (1983). Generation of spatially periodic patterns by a mechanical instability: a mechanical alternative to the Turing model. Report, Department of Biology, University of North Carolina, Chapel Hill, NC 27514.

Hofstadter, D. R. (1981). Strange attractors. *Sci. Am.,* 245:16–29.

Juliani, M. H., and Klein, C. (1978). A biochemical study of the effects of cAMP pulses on aggregateless mutants of *Dictyostelium discoideum. Dev. Biol.,* 62:162–72.

Keller, E. F., and Segel, L. A. (1970). Initiation of slime mold aggregation viewed as an instability. *J. Theoret. Biol.,* 26:399–415.

– (1971). Travelling bands of chemotactic bacteria: a theoretical analysis. *J. Theoret. Biol.,* 30:235–48.

Kemner, W. (1983). A model of head regeneration in *hydra.* Unpublished report. Max Planck Inst. für Medizinische Forshung, Jahstr. 29, 6900 Heidelberg, FRG.

Klein, C. (1976). Adenylate cyclase activity in *Dictyostelium discoideum* amoebae and its changes during differention. *F.E.B.S. Letters,* 68:125–8.

– (1979). A slowly dissociating form of the cell surface cyclic adenosine 3':5'-monophosphate receptor of *Dictyostelium discoideum. J. Biol. Chem.,* 254:12573–8.

Klein, C., and Darmon, M. (1977). Effect of cyclic AMP pulses on adenylate cyclase and the phosphodiesterase inhibitor of *D. discoideum. Nature,* 268:76–8.

Kleppner, D., and Ramsay, N. (1965). *Quick Calculus.* New York: Wiley.

Koshland, D. E., Nemethy, G., and Filmer, D. (1966). Comparison of experimental binding data and theoretical models in proteins containing subunits. *Biochemistry,* 5:365–85.

Lin, C. C., and Segel, L. A. (1974). *Mathematics Applied to Deterministic Problems in the Natural Sciences.* New York: Macmillan.

Loomis, W. F. (1982). *The Development of Dictyostelium discoideum.* New York: Academic Press.

McCarley, R. W., and Hobson, J. A. (1975). Neuronal excitability modulation over the sleep cycle: A structural and mathematical model. *Science,* 189:58–60.

Mackey, M. C., and Glass, L. (1977). Oscillation and chaos in physiological control systems. *Science,* 197:287–9.

MacWilliams, H. K. (1982). Numerical simulations of hydra head regeneration using a proportion-regulating version of the Gierer–Meinhardt model. *J. Theoret. Biol.,* 99:681–703.

Maeda, Y., and Gerisch, G. (1977). Vesicle formation in *Dictyostelium discoideum* cells during oscillations of cAMP synthesis and release. *Exp. Cell Res.,* 110:119–26.

Malchow, D., Nägele, B., Schwarz, H., and Gerisch, G. (1972). Membrane-bound cyclic AMP phosphodiesterase in chemotactically responding cells of *Dictyostelium discoideum. Eur. J. Biochem.,* 28:136–42.

Martiel, J.-L., and Goldbeter, A. (1981). Metabolic oscillations in biochemical systems controlled by covalent enzyme modification. *Biochimie,* 63:119–24.

May, R. M. (1974). Biological populations with nonoverlapping generations: Stable points, stable cycles, chaos. *Science,* 186:645–7.

– (1975). Biological populations obeying difference equations: Stable points, stable cycles, and chaos. *J. Theoret. Biol.,* 51:511–24.

– (1979). Bifurcations and dynamic complexity in ecological systems. *Ann. N.Y. Acad. Sci.,* 316:517–29.

Maynard Smith, J. (1968). *Mathematical Ideas in Biology.* Cambridge University Press.

– (1978). Optimization theory in evolution. *Ann. Rev. Ecol. Systems,* 9:31–56.

Meinhardt, H. (1974). The formation of morphogenetic gradients and fields. *Ber. Deutsch. Bot. Ges.,* 87:101–8.

– (1982). *Models of Biological Pattern Formation.* London: Academic Press.

Meinhardt, H., and Gierer, A. (1974). Applications of a theory of biological pattern formation based on lateral inhibition. *J. Cell Sci.,* 15:321–46.

Michaelis, L., and Menten, M. L. (1913). Die Kinetik der Invertinwirkung. *Biochem. Z.,* 49:333–69.

Mitchison, G. J. (1977). Phyllotaxis and the Fibonacci series. *Science,* 196:270–5.

– (1980). A model for vein formation in higher plants. *Proc. R. Soc. Lond. [B],* 207:79–109.

– (1981). The polar transport of auxin and vein patterns in plants. *Philos. Trans. R. Soc. Lond. [B],* 295:461–71.

Monod, J. (1950). La technique de culture continue; theorie et applications. *Ann. Inst. Pasteur,* 79:390–401.

Monod, J., Wyman, J., and Changeux, J.-P. (1965). On the nature of allosteric transition: A plausible model. *J. Mol. Biol.,* 12:88–118.

Murray, J. D., Oster, G. F., and Harris, A. K. (1983). A mechanical model for mesenchymal morphogenesis. *J. Math. Biol.,* 17:125–9.

Newby, J. C. (1980). *Mathematics for the Biological Sciences – from Graphs through Calculus to Differential Equations.* Oxford: Clarendon Press.

Newell, P. C., and Ross, F. M. (1982). Inhibition by adenosine of aggregation centre initiation and cyclic AMP binding in *Dictyostelium. J. Gen. Microbiol.,* 128:2715–24.

Nisbet, R. M., and Gurney, W. S. C. (1982). *Modelling Fluctuating Populations.* New York: Wiley.

Novick, A., and Szilard, L. (1950). Experiments with the chemostat on spontaneous mutation of bacteria. *Proc. Natl. Acad. Sci. U.S.A.,* 36:708–19.

Odell, G., Oster, G., Alberch, P., and Burnside, B. (1981). The mechanical basis of morphogenesis. I. Epithelial folding and invagination. *Dev. Biol.,* 85:446–62.

Oster, G. F., Murray, J. D., and Harris, A. K. (1984). Mechanical aspects of mesenchymal morphogenesis. *J. Embryol. Exp. Morph., (in press).*

Oster, G. F., and Odell, G. M. (1984). The mechanochemistry of cytogels. In: *Fronts, Interfaces and Patterns,* edited by A. Bishop. Amsterdam: North Holland.

Oster, G., Odell, G., and Alberch, P. (1980). Mechanics, morphogenesis and evolution. In: *Lectures on Mathematics in the Life Sciences 13: Some Mathematical Questions in Biology,* edited by G. Oster, pp. 165–255. Providence: American Mathematical Society.

Oster, G. F., and Wilson, E. O. (1978). *Caste and Ecology in the Social Insects.* Princeton University Press.

Parnas, H., and Cohen, D. (1976). The optimal strategy for the metabolism of reserve materials in micro-organisms. *J. Theoret. Biol.,* 56:19–55.

Parnas, H., and Segel, L. A. (1980). A theoretical explanation for some effects of calcium on the facilitation of neurotransmitter release. *J. Theoret. Biol.,* 84:3–29.

Richter, P. H., and Schranner, R. (1978). Leaf arrangement. Geometry, morphogenesis, and classification. *Naturwissenschaften,* 65:319–27.

Richtmayer, R. D., and Morton, K. W. (1967). *Difference Methods for Initial-Value Problems,* 2nd ed. New York: Interscience.

Riedel, V., and Gerisch, G. (1971). Regulation of extracellular cyclic-AMP phosphodiesterase activity during development of *Dictyostelium discoideum. Biochem. Biophys. Res. Commun.,* 42:119–24.

Roughgarden, J. (1979). *Theory of Population Genetics and Evolutionary Ecology: An Introduction.* New York: Macmillan.

Rubinow, S. I. (1975). *Introduction to Mathematical Biology.* New York: Wiley.

Sachs, T. (1969). Polarity and the induction of organized vascular tissues. *Ann. Bot.,* 33:263–75.

Schaller, H. C., and Bodenmüller, H. (1982). Neurohormones and their functions in hydra. In: *Neurosecretion: Molecules, Cells, Systems,* edited by D. S. Farner and K. Lederis, pp. 381–90. London: Plenum.

Scrive, M., Guespin-Michel, J. F., and Felenbok, B. (1977). Alteration of phosphodiesterase activity after 5-bromodeoxyuridine (BUdR) treatment of *Dictyostelium discoideum. Exp. Cell. Res.,* 108:107–10.

Segel, L. A. (1966). Nonlinear hydrodynamic stability theory and application to thermal convection and curved flow. In: *Non-equilibrium Thermodynamics: Variational Techniques and Stability,* edited by R. J. Donnelly, I. Prigogine, and R. Herman, pp. 165–97. University of Chicago Press.

– (editor). (1980). *Mathematical Models in Molecular and Cellular Biology.* Cambridge University Press.

Segel, L. A., Chet, I., and Henis, Y. (1977). A simple quantitative assay for bacterial motility. *J. Gen. Microbiol.,* 98:329–37.

Shaffer, B. M. (1962). *The acrasina. Adv. Morphogen.,* 2:109–82.

Slobodkin, L. (1961). *The Growth and Regulation of Animal Numbers.* New York: Holt, Rinehart, and Winston.

Smith, H. L. (1981). Competitive coexistence in an oscillating chemostat. *S.I.A.M. J. Appl. Math.,* 40:498–522.

Sussman, M., and Schindler, J. (1978). A possible mechanism of morphogenetic regulation in *Dictyostelium discoideum. Differentiation,* 10:1–5.

Townsend, C. R., and Calow, P. (1981). *Physiological Ecology: An Evolutionary Approach to Resource Use.* Sunderland, Mass.: Sinauer.

Turing, A. M. (1952). The chemical basis for morphogenesis. *Philos. Trans. R. Soc. Lond. [B],* 237:37–72.

Wigglesworth, V. B. (1954). Growth and regeneration in the tracheal system of an insect *Rhodnius prolixus* (Hemiptera). *Quart. J. Micr. Sci.*, 95:115–37.

Wilby, O. K., and Webster, G. (1970). Experimental studies on axial polarity in hydra. *J. Embryol. Exp. Morph.*, 24:595–613.

Winfree, A. T. (1980). *The Geometry of Biological Time.* Berlin: Springer-Verlag.

Wolpert, L., Hicklin, J., and Hornbruch, A. (1971). Positional information and pattern regulation in regeneration of hydra. *Symp. Soc. Exp. Biol.*, 25:391–415.

Wolpert, L. Hornbruch, A., and Clarke, M. R. B. (1974). Positional information and positional signalling in Hydra. *Amer. Zool.*, 14:647–63.

Yeh, R. P., Chan, F. K., and Coukell, M. B. (1978). Independent regulation of the extracellular cyclic AMP phosphodiesterase-inhibitor system and membrane differentiation by exogenous cyclic AMP in *Dictyostelium discoideum. Devel. Biol.*, 66:361–74.

Index